MA

$\mathscr{L}$ Transform Theory and Applications

Robert Vích

Institute of Radio Engineering and Electronics,
Czechoslovak Academy of Sciences, Prague, Czechoslovakia

$\mathscr{Z}$ Transform Theory and Applications

D. Reidel Publishing Company

A MEMBER OF THE KLUWER ACADEMIC PUBLISHERS GROUP

Dordrecht / Boston / Lancaster / Tokyo

Library of Congress Cataloging-in-Publication Data

Vích, Robert
$\mathscr{Z}$ transform theory and applications.

(Mathematics and its applications (East European series))
Translation of: Transformace $\mathscr{Z}$ a některá její použití.
Bibliography: p.
Includes index.
1. $\mathscr{Z}$ transformations. I. Title. II. Series: Mathematics and its applications (D. Reidel Publishing Company). East European series.
QA432.V4813 1987 515.7'23 86-26235
ISBN 90-277-1917-9 (D. Reidel)

Translated from the Czech by Michal Basch.

Published by SNTL – Publishers of Technical Literature, Prague,
in co-edition with D. Reidel Publishing Company, P.O. Box 17,
3300 AA Dordrecht, Holland.

Sold and distributed in the U.S.A. and Canada
by Kluwer Academic Publishers,
101 Philip Drive, Norwell, MA 02061, U.S.A.

Sold and distributed in Albania, Bulgaria, Chinese People's Republic, Cuba, Czechoslovakia, German Democratic Republic, Hungary, Korean People's Democratic Republic, Mongolia, Poland, Rumania, the U.S.S.R., Vietnam, and Yugoslavia by ARTIA, Prague.

In all remaining countries
Kluwer Academic Publishers Group,
P.O. Box 322, 3300 AH Dordrecht, Holland.

Printed in Czechoslovakia by SNTL, Prague.

Table of Contents

Series Editor's Preface

Approach your problem from the right end and begin with the answers. Then one day, perhaps you will find the final question.

The Hermit Clad in Crane Feathers in R. van Gulik's The Chinese Maze Murders.

It isn't that they can't see the solution. It is that they can't see the problem.

G. K. Chesterton. The Scandal of Father Brown. The Point of a Pin.

Growing specialization and diversification have brought a host of monographs and textbooks on increasingly specialized topics. However, the "tree" of knowledge of mathematics and related fields does not grow only by putting forth new branches. It also happens, quite often in fact, that branches which were thought to be completely disparate are suddenly seen to be related.

Further, the kind and level of sophistication of mathematics applied in various sciences has changed drastically in recent years: measure theory is used (non-trivially) in regional and theoretical economics; algebraic geometry interacts with physics; the Minkowsky lemma, coding theory and the structure of water meets one another in packing and covering theory; quantum fields, crystal defects and mathematical programming profit from homotopy theory; Lie algebras are relevant to filtering; and prediction and electrical engineering can use Stein spaces. And in addition to this there are such new emerging subdisciplines as "experimental mathematics", "CFD", "completely integrable systems", "chaos, synergetics and large-scale order", which are almost impossible to fit into the existing classification schemes. They draw upon widely different sections of mathematics.

This programme, Mathematics and Its Applications, is devoted to new emerging (sub)disciplines and to such (new) interrelations as exempla gratia:

- a central concept which plays an important role in several different mathematical and/or scientific specialized areas;

- new applications of the results and ideas from one area of scientific endeavour into another;,
- influences which the results, problems and concepts of one field of enquiry have and have had on the development of another.

The Mathematics and Its Application programme tries to make available a careful selection of books which fit the philosophy outlined above. With such books, which are stimulating rather than definitive, intriguing rather than encyclopaedic, we hope to contribute something towards better communication among the practitioners in diversified fields.

Because of the wealth of scholarly research being undertaken in the Soviet Union, Eastern Europe, and Japan, it was decided to devote special attention to the work emanating from these particular regions.

Thus is was decided to start three regional series under the umbrella of the main MIA programme.

A large part of the application of mathematics in such fields as signal processing, electrical engineering and control and aerospace engineering centers around the Laplace transform. But this is a gadget belonging to continuous time systems and thus not so appropriate in the world of microprocessor control and data processing. Its discrete analogue is the $\mathscr{Z}$ transform, the subject matter of this book in the MIA (Eastern Europe) series. Besides, a nice coherent body of theory of the $\mathscr{Z}$ transform is especially distinguished by its wide applicability (to quite varied topics) and that it is a very accessible tool also for those who have not trained as mathematicians. This book testifies to these facts.

The unreasonable effectiveness of mathematics in science ... Eugene Wigner

Well, if you knows of a better 'ole, go to it.
Bruce Bairnsfather

What is now proved was once only imagined. William Blake

As long as algebra and geometry proceeded along separate paths, their advance was slow and their applications limited.
But when these sciences joined company they drew from each other fresh vitality and thenceforward marched on at a rapid pace towards perfection.
Joseph Louis Lagrange

Bussum, April 1986 **Michiel Hazewinkel**

Preface

It has been the objective of the author to write this book so that it serve readers interested in methods of discrete signal processing from various fields. The book is based on the consistent application of the theory of functions of the complex variable and it assumes a knowledge of the Laplace transform. If the reader is interested only in the theory of the $\mathscr{Z}$ transform, he may restrict himself to reading Chapter 2 and the proofs of the theorems which are given in Section 7.4.

The arrangement of the part of the book devoted to applications follows the author's belief that the majority of those interested in discrete signal processing methods have ample experiences in continuous signal processing, perhaps also in the analysis and synthesis of continuous systems. Therefore, we point out the analogies of the procedures employed in the discrete domain with those employed in the continuous domain.

This book results from a substantial revision and extension of the author's publication of 1964 [46]. The book covers knowledge acquired from problems solved for a number of industrial enterprises and research institutes. All the problems were solved at the Institute of Radio Engineering and Electronics of the Czechoslovak Academy of Sciences not only theoretically but also by computer simulation and, in many cases, they were directly implemented in the construction of the respective devices.

To the experienced user of discrete signal processing methods many of the selected applications and examples may possibly seem too simple. However, they were chosen on the basis of experience gained by the author from teaching in graduate level courses organized by the Faculty of Electrical Engineering of the Technical University in Prague, and in education courses for various industrial enterprises.

Besides these simple examples the reader will also find mentioned in the book procedures which may stimulate him/her to creatively exploit new applications. At present, it seems that it is rather important to find new applications of discrete methods by the cooperation of specialists from diverse fields, in which the advantages of digital processing would be freely and sensitively exploited. The solution of actual requirements will then have favourable influence on the further development of discrete system theory.

The author wishes to thank the Institute of Radio Engineering and Electronics of the Czechoslovak Academy of Sciences for their interest in the advancement of this range of enquiries. The author values the creative cooperation of his colleagues in the research team, among them also his wife. She has not only programmed all the problems discussed, but has also assisted in writing this book.

Robert Vích

Prague, September 1984

Chapter 1

Discrete Signals and Systems

1.1. Introduction

Processing of *signals* is encountered when solving diverse problems arising not only in the engineering fields and in the natural sciences, but also in medicine, economics, and statistics. By signal processing we understand the transformation of a given signal into another signal with required properties. As a rule, a signal is represented as a function of time, although multidimensional signals are also often encountered. These are, moreover, functions of space variables. For instance, such is the case of image processing or of the processing of periodic structures. We represent the processing of a signal by a symbolical block diagram as indicated in Fig. 1. The function $f(t)$ is called the *input signal*, the function $h(t)$ is called the *output signal*. The system G transforms the input signal $f(t)$ to the required output signal $h(t)$.

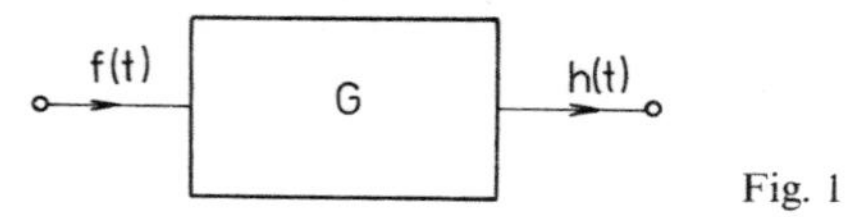

Fig. 1

Signals are divided into *continuous signals* which are functions of a continuous variable (e.g., time), and *discrete signals* which are functions of a discrete variable. A discrete variable takes on only values which are integer

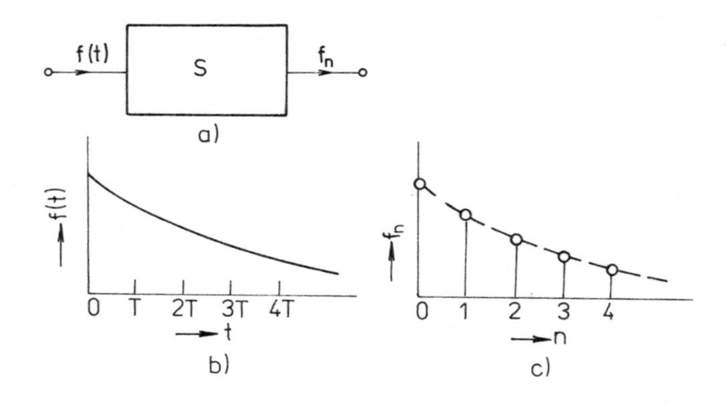

Fig. 2

multiples of a constant basic value. Thus, in fact, discrete signals are *sequences.* In some applications the discrete character of a signal is implied by the physical or mathematical formulation of the discussed problem (e.g., pulse radar), in other applications it is the consequence of the discretization of a continuous signal. The discretization of a signal is indicated in Fig. 2, where the block S can be represented as a switch which generates from the function $f(t)$ a sequence of its values f_n by periodic switching. It is obvious that discretization is a process of the generation of a sequence of function values of a uniquely defined function $f(t)$ of the continuous variable t, $-\infty < t < +\infty$, which assumes the values $t = nT$, $n = \ldots, -2, -1, 0, 1, 2, \ldots$, $T > 0$. Here, T is the *discretization step.* Thus, we have $f_n = f(t)|_{t=nT} = f(nT)$. In what follows we shall use the notation $\{f_n\} = \{\ldots, f_{-2}, f_{-1}, f_0, f_1, f_2, \ldots\}$ for the obtained sequence. In technical applications the term *sampling* is generally used for discretization. The value T is called the *sampling period* or the *length of the sampling interval.* Its reciprocal value $f_s = 1/T$ is called the *sampling frequency.* Often we use the concept of *radian sampling frequency*, even for discrete signals; it is defined by the formula $\omega_s = 2\pi f_s = 2\pi/T$.

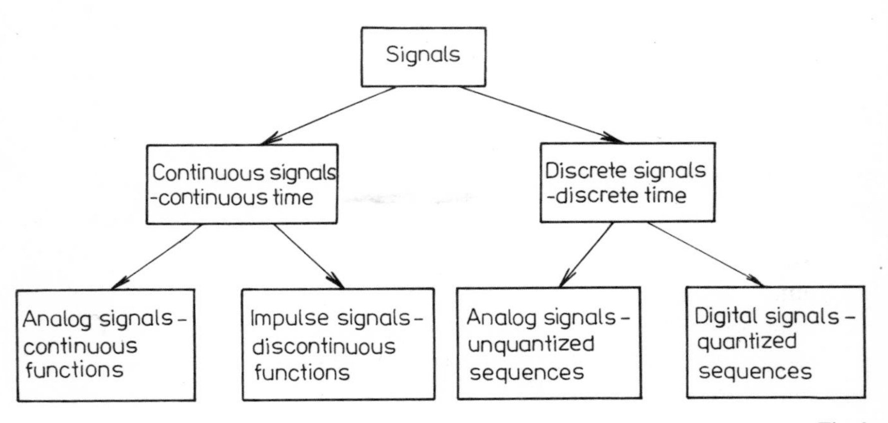

Fig. 3

In the literature continuous signals are often identified with analog signals, and *vice versa.* On the other hand, discrete signals are often interpreted as digital signals. The difference between the individual concepts is obvious from the diagram in Fig. 3.

According to the types of signals the *systems* which process them are also divided into *continuous* and *discrete*. Thus, we speak of systems which work continuously when the input and output signals are continuous functions of time, and of discrete systems when the input and output signals are sequences. It is obvious that hybrid systems may also exist; and it is perhaps possible to realize one system by the other with the aid of suitable transformation elements. For instance, if we desire to use discrete signal processing in a system on whose input there is a continuous signal and on the output of which a continuous signal is required again, we have to use a block diagram according to Fig. 4. In this system the input signal $f(t)$ of the continuous variable t is converted by sampling (which is carried out in block S) with sampling period T to a sequence $\{f_n\}$ which is transformed by block G into the sequence $\{h_n\}$ with the required properties. With the aid of the interpolation block I we then reconstruct, from the sequence $\{h_n\}$, the output signal $\tilde{h}(t)$ which approximates the function $h(t)$ of Fig. 1 in a prescribed manner. It is obvious that the chain of three blocks can be represented by a single block $\tilde{G}$ which replaces in a certain approximate manner the block G of Fig. 1.

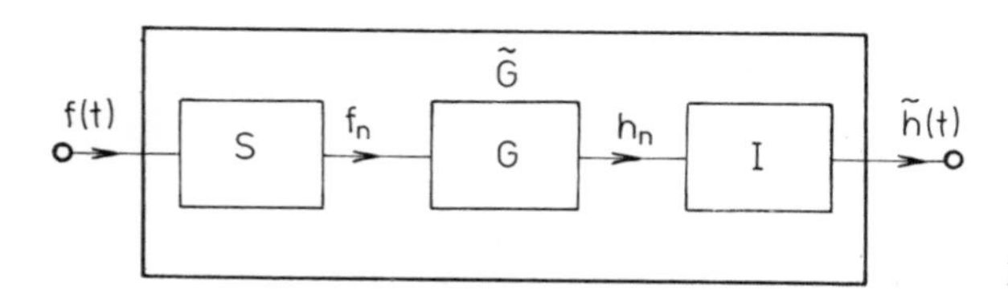

Fig. 4

Problems of this type date from far back, and they are generally used in numerical calculus. As an example it is possible to mention algorithms of numerical differentiation and integration, eventually interpolation. However, it was only in the recent decades that this principle began to be exploited in discrete control systems, and with the development and application of digital computers it came to be of use also in problems traditionally realized by continuous systems. The possibility of the easy programming of problems which arise from the most diverse fields, the considerable accuracy of the calculations, the introduction of even physically unrealizable conditions and signals, made possible the simulation of systems operating continuously. The rapid development of computer technology led to the construction of single purpose digital computers, which make possible in conjunction with analog-digital and digital-analog converters the replacement in real time of continuously operating systems or filters by discrete digital systems or filters. *Digital filters* are devices which are designed to be capable, with minimal necessary precision of the mathematical operations needed for the implementation of the prescribed algorithms, of processing in real time of signals of frequencies up to hundreds of kilohertz. It is obvious that similar devices are of advantage even for the processing of extremely slow signals, for which the currently used analog systems designed for their processing are disproportionately complicated and cumbersome. Let us mention only some of the fields in which digital filters are currently used: in measuring technique they are used for the investigation of vibrations, in spectral and harmonic analysers; for the analysis and synthesis of speech; in geophysics it is for the analysis of earthquakes; in medicine for ECG and EEG analysis; in image processing and pattern recognition; in communications for data transmission and the transformation of signals from frequency division systems to systems with time division of signals; in radiolocation for the filtration of signal

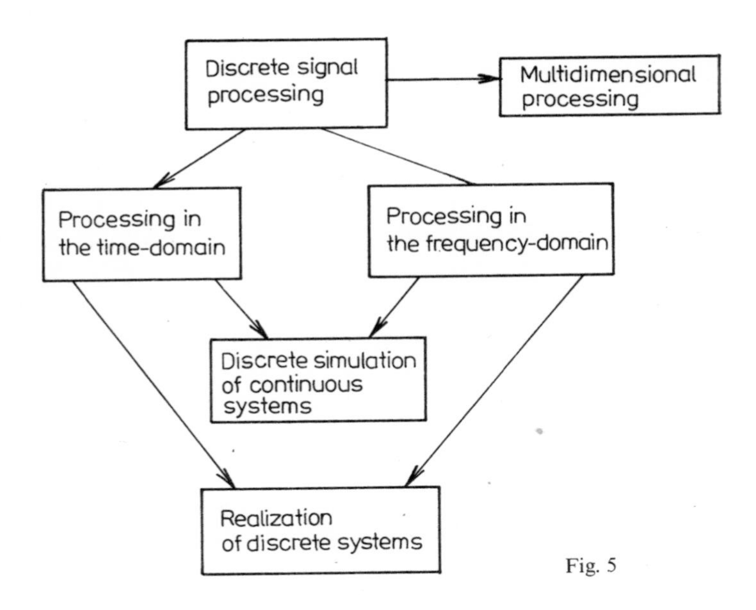

Fig. 5

from noise and to suppress reflections from immobile targets, for telemetric signal processing; etc.

In the individual practical applications it is possible to design discrete systems in dependence on the criteria which are determinant for the considered problems. For some problems the properties of the output signal in the time-domain are essential, for other problems it is in the frequency-domain. One of the possible approaches to the classification of viewpoints for discrete signal processing is indicated in Fig. 5.

1.2. $\mathscr{Z}$ Transform

From the discussion in the previous section it emerges that a discrete signal is formed by a *sequence*. The general *n*th term of the sequence will be denoted by the symbol f_n, $n = \ldots, -2, -1, 0, 1, 2, \ldots$. For the sequence we use the notation

$$\{f_n\} = \{\ldots, f_{-2}, f_{-1}, f_0, f_1, f_2, \ldots\}.$$

Similarly as we describe the operation of continuous systems by differential equations, the operation of discrete systems is represented by *difference equations*. Difference equations are solved either by methods of difference calculus or with the aid of *functional transformations of sequences*. These transformations represent a tool analogous, for instance, to the Laplace transform which is used for the analysis of continuous systems. They make possible an easy transition from the space of object functions represented mostly as functions of time to the space of transforms represented as functions of a complex variable or eventually of frequency. The *$\mathscr{Z}$ transform* is the most widely used of the functional transformations of sequences in the theory of discrete signals and systems. However, even other functional transformations are being used, e.g. the discrete Laplace transform or the discrete Fourier transform. The $\mathscr{Z}$ transform is defined as a functional transformation of sequences. The *sequence* $\{f_n\}$ whose terms may be complex, in general, will mostly be assumed to be *one-sided*, i.e. we assume that the condition $f_n = 0$ for $n < 0$ is satisfied. Section 2.4 will be devoted to general, two-sided sequences, and also to the two-sided $\mathscr{Z}$ transform.

DEFINITION. The *$\mathscr{Z}$ transform* is defined by the relation

$$F(z) = \sum_{n=0}^{+\infty} f_n z^{-n} \tag{1.1}$$

under the assumption that the series is convergent. A sequence is called *$\mathscr{Z}$ transformable* if the series (1.1) converges for at least one (finite) complex z.

Similarly as in the case of other functional transformations we call the sequence $\{f_n\}$ the *object function*, while the function $F(z)$ is called the *transform*. The object function is denoted always by a lowercase letter with subscript n, the transform is denoted by the corresponding capital letter.

For the $\mathscr{Z}$ transform of a sequence $\{f_n\}$ we shall currently use the symbolic notation

$$\mathscr{Z}\{f_n\} = F(z).$$

Before discussing some of the fundamental properties of the transform we demonstrate its application on several basic sequences:

1. Let a sequence $\{f_n\}$ be given whose terms are identically equal to zero with the exception of the term for $n = 0$. Thus, we have $f_0 = 1$, $f_n = 0$ for $n \neq 0$. Upon substitution of these conditions into the definitorical formula the infinite series reduces to a single term and we have

$$\mathscr{Z}\{f_n\} = 1.$$

A sequence of this type plays a similar role in discrete signals and systems as the Dirac function $\delta(t)$ plays in continuous system theory. The considered sequence is sometimes called the *discrete δ function*, or the unit sample. It is indicated in Fig. 6a.

2. Let the finite sequence $\{f_n\} = \{f_0, f_1, \ldots, f_{N-1}\}$, $N < +\infty$, be given, i.e. we have $f_n = 0$ for $N \leqq n$. The definitorical formula reduces to a finite sum which converges always when $|f_n| < +\infty$, $n = 0, 1, \ldots, N-1$:

$$\mathscr{Z}\{f_n\} = \sum_{n=0}^{N-1} f_n z^{-n}.$$

The transform of this sequence is a polynomial in the variable z^{-1} which can be modified to a rational function upon factoring out z^{-N+1}:

$$\sum_{n=0}^{N-1} f_n z^{-n} = \frac{f_0 z^{N-1} + f_1 z^{N-2} + \ldots + f_{N-2} z + f_{N-1}}{z^{N-1}}.$$

If the relation $f_n = c^n$ holds for the nth term of the sequence for $n \leqq N-1$, we have a finite geometric series for whose sum there exists the formula

$$\sum_{n=0}^{N-1} c^n z^{-n} = \sum_{n=0}^{N-1} \left(\frac{c}{z}\right)^n = \frac{1-\left(\frac{c}{z}\right)^N}{1-\frac{c}{z}} = \frac{z^N - c^N}{z^{N-1}(z-c)}. \tag{1.2}$$

In Fig. 6b we find a finite geometric sequence for $c > 1$ and $N = 5$.

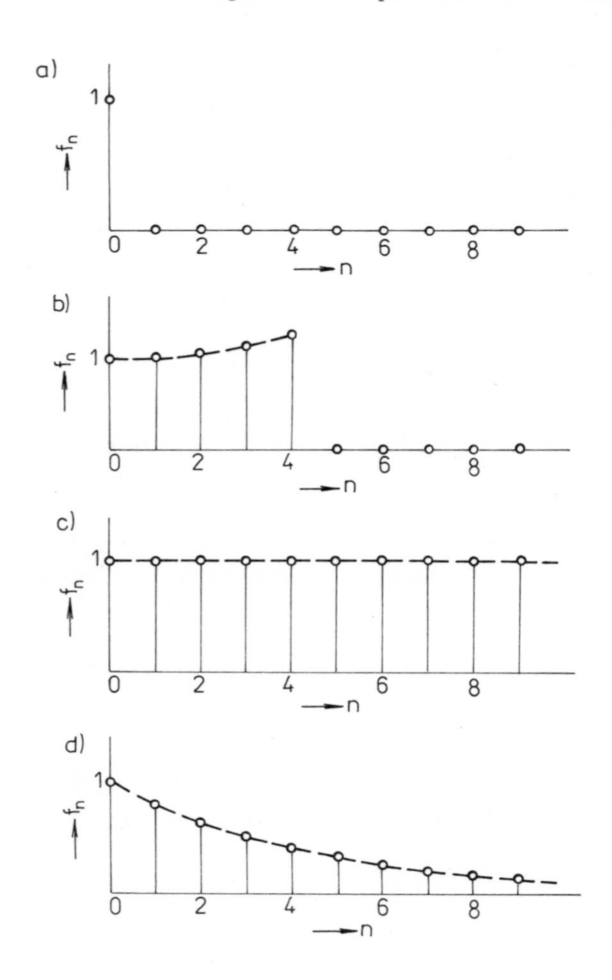

Fig. 6

3. Another important sequence in discrete signals and systems is the unit sequence which may be called the *discrete unit step*. It is defined by the formula (see Fig. 6c)

$$f_n = 1 \qquad \text{for} \qquad n \geqq 0 .$$

Upon substitution into (1.1) we obtain an infinite geometric series which converges for $|z| > 1$ and can be represented by

$$\mathscr{Z}\{f_n\} = \sum_{n=0}^{+\infty} z^{-n} = \frac{1}{1 - z^{-1}} = \frac{z}{z - 1}. \tag{1.3}$$

4. As the last example of a simple sequence let us present the geometric sequence $\{f_n\} = \{e^{an}\}$. The $\mathscr{Z}$ transform yields again an infinite geometric series

$$\mathscr{Z}\{f_n\} = \sum_{n=0}^{+\infty} e^{an} z^{-n} = \sum_{n=0}^{+\infty} \left(\frac{e^a}{z}\right)^n$$

which converges for $|e^a/z| < 1$, i.e. for $|z| > |e^a|$. The transform can be represented by the following rational function of the variable z:

$$\mathscr{Z}\{e^{an}\} = \frac{1}{1 - \dfrac{e^a}{z}} = \frac{z}{z - e^a}. \tag{1.4}$$

Fig. 6d shows the real exponential sequence for $e^a < 1$, i.e. for $a < 0$.

The examples given above are to illustrate the application of the $\mathscr{Z}$ transform on simple but important sequences. Basic knowledge on power series was sufficient here. To comprehend more profoundly the substance of the $\mathscr{Z}$ transform and to complete the preparations for the understanding and application of its complete apparatus, we concentrate our attention below to basic properties of transforms. The reader may find the proofs of the theorems in Appendix 7.4. Since the transforms of sequences are given by Laurent series, namely by their main parts, it is possible to prove most of the theorems with the application of the theory of Laurent series, or power series of complex variables.

The first theorem determines the region of convergence of the series in the definition of the $\mathscr{Z}$ transform.

THEOREM 1. *Let the series* (1.1) *converge at the point* $z_0 \neq 0$. *Then the series converges absolutely outside the circle* $|z| > |z_0|$, *and in every closed region* $|z| \geqq R' > |z_0|$ *it converges uniformly.*

The theorem implies that to every sequence $\{f_n\}$ there exists a number R with the following properties:

The series $\sum_{n=0}^{+\infty} f_n z^{-n}$ converges for all $|z| > 1/R$ and diverges for all $|z| < 1/R$. If the series converges for all $|z| > 0$, we write $R = +\infty$. If the series diverges for all $|z| > 0$, we write $R = 0$. The number R is defined as the least upper bound of the values $|z_0|^{-1}$ and we call it the *radius of convergence* at the point $z = \infty$ of the transform of the sequence $\{f_n\}$.

It is obvious that $1/R = r$ is the radius of a circular region with centre at the origin. For all z which lie inside this region the definition series diverges, while it converges for all z which lie outside this region.

In some applications, complex functions of the real variable ξ which lies in the interval $[a, b]$ are terms of the sequence. Thus, the sequence is of the form $\{f_n(\xi)\}$. Before applying the $\mathscr{Z}$ transform to such sequences it is necessary to verify the uniform convergence of the defining series. To this aim we use the following theorem:

THEOREM 2. *The series* $\sum_{n=0}^{+\infty} f_n(\xi)\, z^{-n}$ *converges uniformly for* ξ *which lie in the interval* $[a, b]$ *if there exist numbers* $z_0 \neq 0$ *and* $M > 0$ *such that*

$$|f_n(\xi)| < M|z_0|^n$$

holds for all $n \geqq n_0 \geqq 0$ *and for all* ξ *from the interval* $[a, b]$.

If we put $\xi = \text{const.}$, we obtain an upper bound criterion for the convergence of the series. Let us define a *sequence of exponential type* as a sequence for which it is possible to find such numbers $M > 0$, $s_0 \geqq 0$, $n_0 \geqq 0$ that the relation

$$|f_n| < M e^{s_0 n}$$

holds for all $n \geqq n_0$. Then it is obvious that the following theorem is valid:

THEOREM 3a. *Every sequence of exponential type is* $\mathscr{Z}$ *transformable.*

Although the next theorem is related to Theorem 19 on the inverse $\mathscr{Z}$ transform, it is reasonable to present it at this stage:

THEOREM 3b. *For a sequence* $\{f_n\}$ *to be* $\mathscr{Z}$ *transformable it is necessary (and by Theorems* 2 *and* 3a *sufficient) that it be of the exponential type.*

An outstanding property of the $\mathscr{Z}$ transform is its uniqueness which can be summed up into the following two theorems:

THEOREM 4a. *If the series* $\sum_{n=0}^{+\infty} f_n z^{-n}$ *converges for* $|z| > 1/R$, *then* $F(z)$ *is a regular function*[1]) *in this region and it is the unique transform of the sequence* $\{f_n\}$.

THEOREM 4b. *Let* $F(z)$ *be a regular function in the region* $|z| > 1/R$. *Then there exists a single sequence* $\{f_n\}$ *for which* $\mathscr{Z}\{f_n\} = F(z)$.

Hence, it follows that any arbitrarily obtained expansion of the function $F(z)$, which is regular in the circle with centre at the point $z = \infty$ and radius $1/R$, into the series $\sum_{n=0}^{+\infty} c_n z^{-n}$ leads to a sequence of coefficients $\{c_n\}$ for which we have $c_n = f_n$, $n = 0, 1, 2, \ldots$. For this reason the transform $F(z)$ is sometimes called, in the literature, the "generating function" of the sequence $\{f_n\}$. This useful property of the $\mathscr{Z}$ transform is often used for the inverse $\mathscr{Z}$ transform (see Subsection 2.2.2).

1.3. Methods of the Analysis of Discrete Signals and Systems

In Section 1.1 the concepts of continuous signal and continuous system as well as the interpretation of the discrete signal and discrete system were presented. The behaviour of continuous systems is represented by differential equations (or rather by systems of differential equations), where the input signal of a continuous system is the function which appears on the right-hand side of the differential equation. The solution of the differential equation then yields the response of the continuous system to a given input signal. If the differential equations which describe a given system are linear, we speak of a linear system. Moreover, if the coefficients of the differential equation are independent of time, we have a linear stationary system.

In engineering problems the application of functional transformations became customary for the solution of linear stationary systems. Of these

[1]) In recent textbook literature the concept of "holomorphic function" is being used instead of "regular function".

transformations the *Laplace transform* is at present the most widely used. The exposition found in this book is based on the Laplace transform of the function $f(t)$ defined for $t \geqq 0$ by the relation

$$F_{\mathrm{L}}(p) = \int_0^{+\infty} f(t)\,\mathrm{e}^{-pt}\,\mathrm{d}t\,, \qquad p = \sigma + \mathrm{j}\omega\,, \ ^{2)}$$

for which we shall often use the symbolic notation

$$\mathscr{L}\{f(t)\} = F_{\mathrm{L}}(p)\,.$$

The function $f(t)$ is called the *object function*, the function $F_{\mathrm{L}}(p)$ its *Laplace transform.* The *inverse Laplace transform* is given by the integral

$$f(t) = \frac{1}{2\pi\mathrm{j}} \int_{c-\mathrm{j}\infty}^{c+\mathrm{j}\infty} F_{\mathrm{L}}(p)\,\mathrm{e}^{pt}\,\mathrm{d}p \qquad \text{for} \qquad t > 0\,,$$

for which we shall use the simplified notation

$$f(t) = \mathscr{L}^{-1}\{F_{\mathrm{L}}(p)\}\,.$$

It is natural that in the initial stages of the development of discrete system theory these systems were interpreted as special cases of continuous systems in which discontinuous signals of the continuous variable t are active. This led to the description of their behaviour by difference equations, but these were solved by the Laplace transform. In due time this approach proved to be unnecessarily general, on the one hand; on the other hand, it does not correspond frequently to the physical activity of discrete systems in which sequences appear as signals, not, e.g., staircase functions. For this reason a number of variously defined functional transformations of sequences were introduced. Most probably, the first one to introduce the $\mathscr{Z}$ transform was W. Hurewicz in the book [14] which was published in 1947. To the sequence $\{f_n\}$ Hurewicz assigned the power series $\sum_{n=0}^{+\infty} f_n z^{-n}$ and called it the $\mathscr{Z}$ transform. Later, most authors accepted his approach but defined the $\mathscr{Z}$ transform of the function $f(t)$ as the Laplace transform of an infinite series of translated Dirac functions $\delta(r - nT)$ whose unit

[2]) In mathematical literature the imaginary unit is denoted mostly by the symbol i.

area is modulated by the function $f(t)$ [15]. It was necessary to define the Laplace transform on the basis of the Stieltjes integral

$$\mathscr{L}\left\{\sum_{n=0}^{+\infty} f(t)\,\delta(t-nT)\right\} = \sum_{n=0}^{+\infty} f(nT)\,\mathrm{e}^{-npT} = F(\mathrm{e}^{pT}).$$

In the resulting Laplace transform the substitution $\mathrm{e}^{pT} = z$ was then introduced.

In the Soviet Union it was Ya. Z. Tzypkin [44] who introduced the *discrete Laplace transform* by the relation

$$F(\mathrm{e}^{p}) = \sum_{n=0}^{+\infty} f(n)\,\mathrm{e}^{-pn}$$

in a number of his papers of 1949. Some authors call the thus defined transform the *Dirichlet transform* since the series which defines the discrete Laplace transform is a Dirichlet series, essentially.

We shall not loose time here by the description of the different definitions of the functional transformations of sequences which differ mutually either in the notation of the variable or in the change of the subscript n to $n + 1$ or to $-n$, etc.

In principle, functional transformations of sequences can be divided into two groups:

– transformations defined by a power series,
– transformations defined by a Dirichlet series.

Let us try to compare the two approaches.

The $\mathscr{Z}$ transform is defined by a power series of the variable z^{-1}, thus by a Laurent series. For this reason some authors ([12], [60]) recommend that the transform be called the Laurent transform. (It is rather unusual that the name of a transform be chosen in agreement with the alphabetic symbol which was used.) The Laurent series theory is relatively simple and familiar – as compared with the Dirichlet series theory. For instance, in the case of power series the region of convergence and absolute convergence is identical with the regularity region of the sum of the series, and it is circular. In Dirichlet series theory these regions are halfplanes and the region of convergence, the region of absolute convergence, and the regularity region of the sum of the series may be generally different [3]. Besides, the notation of a power series is simpler than the notation of a Dirichlet series.

The advantage of the discrete Laplace transform can be perceived in the fact that it can be considered a discrete analogy of the Laplace transform. Several theorems are then formally similar. However, it is obvious, e.g., that it is possible to pass easily from one transform to the other when using the list of transforms.

From this review of functional transformations of sequences it follows that the definition of the $\mathscr{Z}$ transform presented in this book corresponds to the Hurewicz approach. In 1964, when the author's book [46] was published based on the thus defined $\mathscr{Z}$ transform, various definitions still appeared in the literature, e.g. even the *discrete Fourier transform* defined by the relation

$$F_{\mathrm{L}}(\mathrm{j}\omega) = \sum_{n=0}^{+\infty} f_n \, \mathrm{e}^{-\mathrm{j}\omega n} .$$

This transform could be considered to be a special case of the discrete Laplace transform for $p = \mathrm{j}\omega$. In due time, the advantages and simplicity of the $\mathscr{Z}$ transform for the representation of aperiodic signals gained ground.

In several physical problems signals occur which are periodic, or rather which can be considered periodic under certain assumptions. For their representation the so-called *finite discrete Fourier transform* was therefore introduced, defined by the relation

$$F_k = \sum_{n=0}^{N-1} f_n \, \mathrm{e}^{-\mathrm{j}nk(2\pi/N)} , \qquad k = 0, 1, \ldots, N-1 .$$

For the inverse finite discrete Fourier transform we then have

$$f_n = \frac{1}{N} \sum_{n=0}^{N-1} F_k \, \mathrm{e}^{\mathrm{j}kn(2\pi/N)} , \qquad n = 0, 1, 2, \ldots, N-1 .$$

However, the discrete Fourier transform (DFT) defined in this way does not serve by far only for the description of signals, but it represents also a very effective method for the computation of the Fourier integral, of the spectra of signals, or signals from their spectra, etc. To the extensive practical exploitation of the discrete Fourier transform the establishment of the algorithm of the so-called *fast Fourier transform* (FFT) contributed considerably. The properties of the discrete Fourier transform are discussed in the literature in special publications (e.g., [10], [29], [31], [32]).

Chapter 2

Properties of the $\mathscr{Z}$ Transform

2.1. Fundamental Theorems on the $\mathscr{Z}$ Transform

In this chapter some properties of the $\mathscr{Z}$ transform will be discussed. The properties will be summarized in theorems, with examples added. Proofs of the theorems are left to the Appendix. The examples help to gain insight into the applications of the individual theorems, and they also serve as the basis for the list of $\mathscr{Z}$ transforms presented in the Appendix.

2.1.1. *Linearity of the Transform*

THEOREM 5. *Let complex numbers* c_i, $i = 0, 1, 2, \ldots, l$, *be given. If there exist transforms of the sequences* $\mathscr{Z}\{f_{i,n}\} = F_i(z)$ *with radii of convergence* $R_i > 0$ *for* $i = 0, 1, 2, \ldots, l$ *(l finite), then there also exists the transform* $\mathscr{Z}\left\{\sum_{i=0}^{l} c_i f_{i,n}\right\}$, *and for* $|z| > \max 1/R_i$ *we have*

$$\mathscr{Z}\left\{\sum_{i=0}^{l} c_i f_{i,n}\right\} = \sum_{i=0}^{l} c_i F_i(z)\,. \tag{2.1}$$

With the aid of this theorem and the examples solved in Chapter 1 further useful transforms of sequences will be determined.

EXAMPLE 1. We wish to determine the transform of the sequences $\{\sinh \alpha n\}$ and $\{\cosh \alpha n\}$. For this we use the familiar relations $\sinh \alpha n = \frac{1}{2}(\mathrm{e}^{\alpha n} - \mathrm{e}^{-\alpha n})$ and $\cosh \alpha n = \frac{1}{2}(\mathrm{e}^{\alpha n} + \mathrm{e}^{-\alpha n})$, together with the theorem on the linearity of the transform. It is obvious that both partial sequences are $\mathscr{Z}$ transformable with convergence radii $R = \min(\mathrm{e}^{\alpha}, \mathrm{e}^{-\alpha})$, i.e. for all $|z| > \max(\mathrm{e}^{\alpha}, \mathrm{e}^{-\alpha})$. With the aid of the transform $\mathscr{Z}\{\mathrm{e}^{an}\}$ from (1.4) we obtain for $a = \alpha$ the relation

$$\mathscr{Z}\{\sinh \alpha n\} = \frac{1}{2}[\mathscr{Z}\{\mathrm{e}^{\alpha n}\} - \mathscr{Z}\{\mathrm{e}^{-\alpha n}\}] =$$

$$= \frac{1}{2}\left[\frac{z}{z - \mathrm{e}^{\alpha}} - \frac{z}{z - \mathrm{e}^{-\alpha}}\right] = \frac{z \sinh \alpha}{z^2 - 2z \cosh \alpha + 1} \tag{2.2}$$

and, analogously,

$$\mathscr{Z}\{\cosh \alpha n\} = \frac{z(z - \cosh \alpha)}{z^2 - 2z \cosh \alpha + 1}. \tag{2.3}$$

Both transforms are rational functions of the variable z which are regular in the entire plane with the exception of the points $z = e^{\alpha}$ and $z = e^{-\alpha}$. For real α these points lie on the real axis.

EXAMPLE 2. Similarly, we determine the transforms of the sequences $\{\sin \beta n\}$ and $\{\cos \beta n\}$. Here, in addition to the relations $\sin \beta n = (1/2j)(e^{j\beta n} - e^{-j\beta n})$ and $\cos \beta n = \frac{1}{2}(e^{j\beta n} + e^{-j\beta n})$, we make use of the transform of the exponential sequence $\mathscr{Z}\{e^{an}\}$ for $a = j\beta$. It is easily verified that both partial sequences are $\mathscr{Z}$ transformable for $|z| > |e^{j\beta}| = 1$, and that we have

$$\mathscr{Z}\{\sin \beta n\} = \frac{z \sin \beta}{z^2 - 2z \cos \beta + 1}, \tag{2.4}$$

$$\mathscr{Z}\{\cos \beta n\} = \frac{z(z - \cos \beta)}{z^2 - 2z \cos \beta + 1}. \tag{2.5}$$

Both transforms are regular in the entire plane except the points $z = e^{j\beta}$ and $z = e^{-j\beta}$ which lie on the unit circle and are complex conjugate.

EXAMPLE 3. The sequence $\{\sin(\beta n + \varphi)\}$ is given. To be able to apply the above results we use the decomposition

$$\sin(\beta n + \varphi) = \sin \beta n \cos \varphi + \cos \beta n \sin \varphi .$$

Under the same convergence conditions we then obtain

$$\mathscr{Z}\{\sin(\beta n + \varphi)\} = \frac{z[z \sin \varphi + \sin(\beta - \varphi)]}{z^2 - 2z \cos \beta + 1}. \tag{2.6}$$

EXAMPLE 4. As the last example which illustrates the application of the linearity theorem, the transform of a sequence selected from the unit sequence is determined. Moreover, this example will be a useful transition to the translation theorem since it will elucidate the significance of the variable z.

Let us have the sequence $\{f_n\}$ given by the formula

$$f_n = 1 \quad \text{for} \quad n = 0, 2, 4, \ldots, 2k, \ldots,$$
$$f_n = 0 \quad \text{for} \quad n = 1, 3, 5, \ldots, 2k + 1, \ldots .$$

For the application of the linearity theorem we express this sequence as the sum of two sequences $\{f_n\} = \{f_{1,n}\} + \{f_{2,n}\}$. For these sequences let us have

$$f_{1,n} = \frac{1}{2},$$

$$f_{2,n} = (-1)^n \frac{1}{2}$$

for $n \geqq 0$. The sequence $\{f_{1,n}\}$ is a unit sequence multiplied by the coefficient $\frac{1}{2}$, its transform is

$$\mathscr{Z}\{f_{1,n}\} = \frac{1}{2} \frac{z}{z-1}.$$

For the second sequence we obtain, after substitution into the definitorical relation and reduction, for $|z| > 1$ the transform

$$\mathscr{Z}\{f_{2,n}\} = \frac{1}{2} \sum_{n=0}^{+\infty} (-1)^n z^{-n} = \frac{1}{2} \sum_{n=0}^{+\infty} (-z)^{-n} = \frac{1}{2} \frac{z}{z+1}.$$

The sum of the transforms is the transform of the desired sequence

$$\mathscr{Z}\{f_n\} = \mathscr{Z}\{f_{1,n}\} + \mathscr{Z}\{f_{2,n}\} = \frac{1}{2}\left[\frac{z}{z-1} + \frac{z}{z+1}\right] = \frac{z^2}{z^2-1}.$$

2.1.2. *Translation of a Sequence*

THEOREM 6. *A positive integer k is given. If the transform $\mathscr{Z}\{f_n\} = F(z)$ exists for $|z| > 1/R$, then there also exist the transform $\mathscr{Z}\{f_{n+k}\}$ and, for $n \geqq k$, the transform $\mathscr{Z}\{f_{n-k}\}$. For $|z| > 1/R$ we have*

$$\mathscr{Z}\{f_{n+k}\} = z^k \left[F(z) - \sum_{n=0}^{k-1} f_n z^{-n}\right], \tag{2.7a}$$

$$\mathscr{Z}\{f_{n-k}\} = z^{-k} F(z). \tag{2.7b}$$

EXAMPLE 1. Let us illustrate the concept of the translation of a sequence and the application of Theorem 6 on the case of the exponential sequence $\{f_n\} = \{e^{\alpha n}\}$. In Fig. 7 we find the original sequence as well as the sequences translated to the left and to the right for $k = 2$. We have to keep in mind that we are working with the one-sided $\mathscr{Z}$ transform, i.e. that before right translation we have $f_n = 0$ for $n < 0$ and after left translation the resulting sequence has to satisfy this condition as well. From Fig. 7 we see that right translation corresponds to the delay of the sequence while left translation may be viewed as its anticipation.

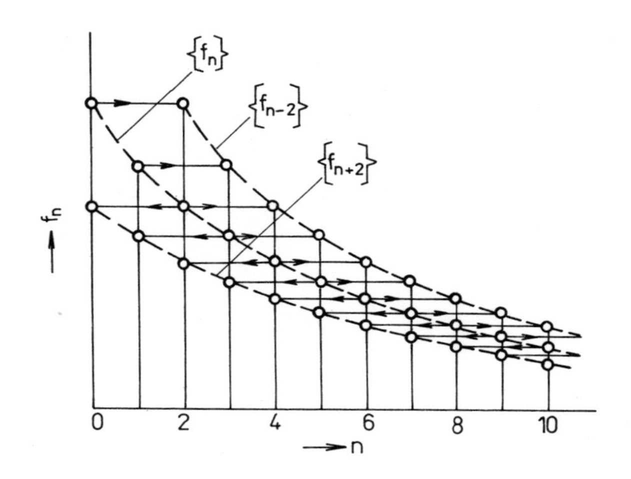

Fig. 7

Since we have

$$\mathscr{Z}\{e^{\alpha n}\} = \frac{z}{z - e^{\alpha}},$$

we obtain, upon application of Theorem 6,

$$\mathscr{Z}\{e^{\alpha(n+2)}\} = z^2[\mathscr{Z}\{e^{\alpha n}\} - (1 + e^{\alpha}z^{-1})] = \frac{e^{2\alpha}z}{z - e^{\alpha}}$$

for $k = 2$ and left translation, and

$$\mathscr{Z}\{e^{(n-2)}\} = z^{-2}\,\mathscr{Z}\{e^{\alpha n}\} = \frac{1}{z(z - e^{\alpha})}$$

for right translation. From the theorem and the example it is obvious that the multiplication of the transform by the variable z leads to left translation, while the division of the transform by the variable z results in right translation.

EXAMPLE 2. As another example of the application of the translation theorem we discuss the construction of a periodic one-sided sequence defined by the relation $\{f_n\} = \{f_{n+N}\}$, where $N > 0$.

If $\mathscr{Z}\{f_n\} = F(z)$ exists, then for the translated sequence we may write, upon application of Theorem 6,

$$F(z) = z^N[F(z) - F_N(z)],$$

where $F_N(z) = \sum_{n=0}^{N-1} f_n z^{-n}$ is the transform of a single period of the given periodic sequence. If we solve the equation for the desired transform of this sequence, we obtain

$$F(z) = \frac{z^N}{z^N - 1} F_N(z). \tag{2.8}$$

In the rational function we easily recognize the sum of the series $\sum_{n=0}^{+\infty} z^{-Nn}$ which is the transform of the unit sequence with step N. Upon substitution into the resulting expression we obtain the transform of a periodic function in expanded form as follows:

$$F(z) = F_N(z) \sum_{n=0}^{+\infty} z^{-Nn}.$$

Since the transform $F_N(z)$ has no other poles than the origin, where it has a pole of multiplicity $N - 1$, it is obvious that the transform of a periodic sequence has poles lying only upon the unit circle, namely at the points which are given by the roots of the equation $z^N - 1 = 0$, i.e. at the points

$$z_i = \mathrm{e}^{\mathrm{j}(2\pi/N)i}, \qquad i = 0, 1, 2, \ldots, N - 1.$$

2.1.3. *Change of Subscript of a Sequence*

THEOREM 7a. *Let the transform of the sequence $\mathscr{Z}\{g_n\} = G(z)$ exist for $|z| > 1/R$, and let i be a positive integer. If $\{f_m\}$ is a sequence defined by the relations*

$$f_m = g_n \quad \text{for} \quad m = ni,$$
$$f_m = 0 \quad \text{for} \quad m \neq ni, \quad m = 0, 1, 2, \ldots,$$

then the transform $\mathscr{Z}\{f_m\}$ *also exists for* $|z| > R^{-1/i}$ *and we have*

$$\mathscr{Z}\{f_m\} = G(z^i). \tag{2.9}$$

THEOREM 7b. *Let the transform of the sequence* $\mathscr{Z}\{f_m\} = F(z)$ *exist for* $|z| > 1/R$, *and let i be a positive integer. If we have*

$$f_m = 0 \qquad \text{for} \qquad m \neq in\,, \quad n = 0, 1, 2, \ldots,$$

for the sequence $\{f_m\}$, *then the function* $F(z^{1/i}) = G(z)$ *is the transform of the sequence* $\{g_n\}$ *for* $|z| > R^{-i}$ *and we have*

$$g_n = f_m \qquad \text{for} \qquad n = \frac{m}{i}.$$

With the aid of Theorem 7a it is possible to solve Example 4 of Subsection 2.1.1, which was used to illustrate the theorem on the linearity of the $\mathscr{Z}$ transform, in a simpler manner.

EXAMPLE. Let the sequence $\{g_n\} = \{e^{an}\}$ be given. The sequence $\{f_m\}$ is given by the formula

$$f_m = e^{an} \qquad \text{for} \qquad m = 0, N, 2N, \ldots, Nn, \ldots,$$
$$f_m = 0 \qquad \text{for} \qquad m \neq Nn, \quad n = 0, 1, 2, \ldots.$$

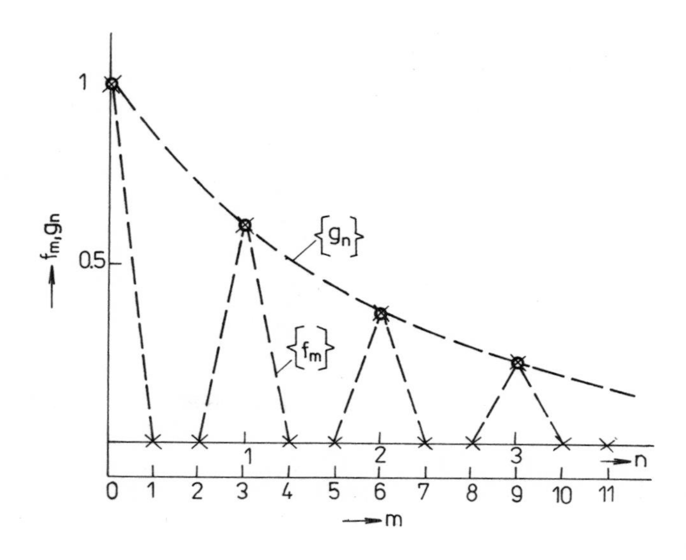

Fig. 8

This means that $N - 1$ zero terms of the sequence $\{f_m\}$ are inserted between every two terms of the sequence $\{g_n\}$. The transform of the sequence $\{e^{an}\}$ is given by relation (1.4) for $|z| > |e^a|$:

$$\mathscr{Z}\{e^{an}\} = \frac{z}{z - e^a}.$$

By Theorem 7a there also exists the transform of the sequence $\{f_m\}$ for $|z| > |e^a|^{1/N}$ and we have

$$\mathscr{Z}\{f_m\} = \frac{z^N}{z^N - e^a}.$$

The two sequences are shown in Fig. 8 for $N = 3$ and $a < 0$.

2.1.4. *Similarity of Transforms*

THEOREM 8. *If the transform* $\mathscr{Z}\{f_n\} = F(z)$ *exists for* $|z| > 1/R$, *and if* $\lambda \neq 0$ *is a complex constant, then the transform* $\mathscr{Z}\{\lambda^n f_n\}$ *also exists and for* $|z| > |\lambda|/R$ *we have*

$$\mathscr{Z}\{\lambda^n f_n\} = F(z/\lambda). \tag{2.10}$$

EXAMPLE. In a number of problems of signal and system theory sequences of the forms $\{e^{\alpha n} \sin \beta n\}$ and $\{e^{\alpha n} \cos \beta n\}$ are encountered. It is possible to determine their transforms with the aid of the linearity theorem, but the application of Theorem 8 is more advantageous.

The transforms of the sequences $\{\sin \beta n\}$ and $\{\cos \beta n\}$ are given by relations (2.4) and (2.5). If we put $\lambda = e^\alpha$, the direct application of Theorem 8 yields

$$\mathscr{Z}\{e^{\alpha n} \sin \beta n\} = \frac{(z/e^\alpha) \sin \beta}{(z/e^\alpha)^2 - 2(z/e^\alpha) \cos \beta + 1} = \frac{z\, e^\alpha \sin \beta}{z^2 - 2z\, e^\alpha \cos \beta + e^{2\alpha}}, \tag{2.11}$$

$$\mathscr{Z}\{e^{\alpha n} \cos \beta n\} = \frac{z(z - e^\alpha \cos \beta)}{z^2 - 2z\, e^\alpha \cos \beta + e^{2\alpha}} \tag{2.12}$$

for $|z| > |e^{\alpha}|$. Both transforms are regular functions in the entire complex plane z with the exception of the points $z_{1,2} = e^{\alpha \pm j\beta}$ at which simple poles lie.

2.1.5. *Convolution of Sequences*

By the *convolution* of two sequences $\{f_n\}$ and $\{g_n\}$ we understand the sequence defined by the relation $\sum_{k=0}^{n} f_k g_{n-k}$. We shall denote it by the symbol $\{f_n\} * \{g_n\} = \{f_n * g_n\}$. For the convolution the following relations hold:

$\{f_n * g_n\} = \{g_n * f_n\}$ – the commutative law,

$\{f_{1,n} + f_{2,n}\} * \{g_n\} = \{f_{1,n} * g_n\} + \{f_{2,n} * g_n\}$ – the distributive law,

$\{f_{1,n} * f_{2,n}\} * \{g_n\} = \{f_{1,n}\} * \{f_{2,n} * g_n\}$ – the associative law.

The following theorem is true:

THEOREM 9. *If there exist the transforms* $\mathscr{Z}\{f_n\} = F(z)$ *for* $|z| > 1/R_1$ *and* $\mathscr{Z}\{g_n\} = G(z)$ *for* $|z| > 1/R_2$, *then the transform* $\mathscr{Z}\{f_n * g_n\}$ *also exists and we have, for* $|z| > \max(1/R_1, 1/R_2)$,

$$\mathscr{Z}\{f_n * g_n\} = F(z) \,.\, G(z) \,. \tag{2.13}$$

The convolution theorem is one of the most important theorems since it yields the means for the computation of the response of discrete systems. Its application will be demonstrated on examples to a certain degree more complicated.

EXAMPLE 1. The sequence $\sum_{k=0}^{n+3} f_k g_{m+3-k}$ is given and we are looking for its transform. Let $\mathscr{Z}\{g_n\} = G(z)$ and $\mathscr{Z}\{f_n\} = F(z)$. If we introduce the sequence

$$h_n = \sum_{k=0}^{n} f_k g_{n-k} ,$$

then the given sequence is the translated sequence $\{h_{n+3}\}$. By Theorem 9 we have

$$\mathscr{Z}\{h_n\} = F(z) \,.\, G(z) \,.$$

With the aid of the translation theorem applied to the sequence $\{h_{n+3}\}$ we obtain

$$\mathscr{Z}\{h_{n+3}\} = z^3 H(z) - z^3 h_0 - z^2 h_1 - z h_2 ,$$

where the initial terms are

$$\begin{aligned} h_0 &= f_0 g_0 , \\ h_1 &= f_0 g_1 + f_1 g_0 , \\ h_2 &= f_0 g_2 + f_1 g_1 + f_2 g_0 . \end{aligned}$$

Upon substitution we obtain the desired transform

$$\mathscr{Z}\left\{\sum_{k=0}^{n+3} f_k g_{n+3-k}\right\} = z^3[F(z)\, G(z) - f_0 g_0] - \\ - z^2(f_0 g_1 + f_1 g_0) - z(f_0 g_2 + f_1 g_1 + f_2 g_0) .$$

EXAMPLE 2. The theorem on the transform of a convolution can be sometimes also applied with advantage to the determination of the object function to a given transform. Let the transform

$$F(z) = \frac{z^2}{(z - e^{\alpha_1})(z - e^{\alpha_2})}$$

be given. From this form it is obvious that the transform originated from the multiplication of the transforms

$$F_1(z) = \frac{z}{z - e^{\alpha_1}} = \mathscr{Z}\{e^{\alpha_1}\} ,$$

$$F_2(z) = \frac{z}{z - e^{\alpha_1}} = \mathscr{Z}\{e^{\alpha_2}\} .$$

However, by the convolution theorem we have

$$F_1(z)\, F_2(z) = \mathscr{Z}\left\{\sum_{k=0}^{n} e^{\alpha_1 k}\, e^{\alpha_2(n-k)}\right\} = \mathscr{Z}\left\{e^{\alpha_2 n} \sum_{k=0}^{n} e^{k(\alpha_1 - \alpha_2)}\right\} .$$

Representing the sum of the finite geometric series on the right-hand side according to the usual formula we obtain the desired sequence

$$\{f_n\} = \left\{e^{\alpha_2 n} \sum_{k=0}^{n} e^{k(\alpha_1 - \alpha_2)}\right\} = \left\{e^{\alpha_2 n} \frac{e^{(\alpha_1 - \alpha_2)(n+1)} - 1}{e^{(\alpha_1 - \alpha_2)} - 1}\right\} .$$

2.1.6. *Transforms of Differences*

The *first difference* of a sequence $\{f_n\}$ is defined by the relation

$$\Delta f_n = f_{n+1} - f_n .$$

The following theorem is true:

THEOREM 10a. *If the transform* $\mathscr{Z}\{f_n\} = F(z)$ *exists for* $|z| > 1/R$, *then the transform* $\mathscr{Z}\{\Delta f_n\}$ *also exists for* $|z| > 1/R$ *and we have*

$$\mathscr{Z}\{\Delta f_n\} = (z - 1)\,F(z) - zf_0 . \tag{2.14}$$

Similarly as the first difference of the sequence $\{f_n\}$ was introduced, we may also introduce the *second difference* as the difference of the first difference:

$$\Delta^2 f_n = \Delta f_{n+1} - \Delta f_n .$$

Generally, we have

$$\Delta^k f_n = \Delta^{k-1} f_{n+1} - \Delta^{k-1} f_n .$$

For the transform of the kth difference we have:

THEOREM 10b. *If the transform* $\mathscr{Z}\{f_n\} = F(z)$ *exists for* $|z| > 1/R$, *then the transform* $\mathscr{Z}\{\Delta^k f_n\}$ *also exists and for* $|z| > 1/R$ *we have*

$$\mathscr{Z}\{\Delta^k f_n\} = (z - 1)^k F(z) - z \sum_{i=0}^{k-1} (z - 1)^{k-i-1} \Delta^i f_0 , \tag{2.15}$$

where $\Delta^i f_0$ *is the ith difference for* $n = 0$ *and* $\Delta^0 f_0 = f_0$.

Relation (2.15) for the transform of the kth difference is exploited with advantage for the construction of the transform of the sequence for which we have $\Delta^k f_n = 0$ for some $k \geqq 1$. If we represent the function $F(z)$ from this relation, we obtain

$$F(z) = \frac{z}{z-1} \sum_{i=0}^{k-1} \frac{\Delta^i f_0}{(z-1)^i} + \frac{1}{(z-1)^k} \mathscr{Z}\{\Delta^k f_n\} . \tag{2.16}$$

EXAMPLE 1. Let us look for the transform of the sequence $\{f_n\} = \{n\}$. We have

$$\mathscr{Z}\{n\} = \sum_{n=0}^{+\infty} nz^{-n} .$$

For this sequence we easily verify that

$$\Delta f_n = (n+1) - n = 1\,,$$
$$\Delta^2 f_n = \Delta f_{n+1} - \Delta f_n = 0\,.$$

All differences of higher order are zero. Upon substitution of these values into relation (2.16) we obtain the desired transform

$$\mathscr{Z}\{n\} = \frac{z}{(z-1)^2}. \tag{2.17}$$

EXAMPLE 2. Let $\{f_n\} = \left\{\binom{n}{2}\right\}$. The enumeration of the differences yields

$$\Delta f_n = \binom{n+1}{2} - \binom{n}{2} = \frac{(n+1)\,n}{2} - \frac{n(n-1)}{2} = n\,,$$
$$\Delta^2 f_n = 1\,.$$

All differences of higher order are again zero. Thus, the transform of the sequence is

$$\mathscr{Z}\left\{\binom{n}{2}\right\} = \frac{z}{(z-1)^3}. \tag{2.18}$$

Analogously, it is possible to verify the correctness of the general relation

$$\mathscr{Z}\left\{\binom{n}{m}\right\} = \frac{z}{(z-1)^{m+1}}. \tag{2.19}$$

2.1.7. *Transform of the Sequence of Partial Sums*

THEOREM 11. *Let the sequence of partial sums* $\sum_{k=0}^{n-1} f_k$ *be given, which is generated from the sequence* $\{f_n\}$. *If* $\mathscr{Z}\{f_n\} = F(z)$ *exists for* $|z| > 1/R$, *then the transform of the sequence of partial sums also exists, and for* $|z| > \max(1/R, 1)$ *we have*

$$\mathscr{Z}\left\{\sum_{k=0}^{n-1} f_k\right\} = \frac{F(z)}{z-1}. \tag{2.20}$$

EXAMPLE 1. The first example elucidates the concept of the sequence of partial sums generated from the finite sequence $\{f_n\} = \{3, -2, 2, -1\}$.

The transform of this sequence is a polynomial in the variable z^{-1}:

$$\mathscr{Z}\{f_n\} = F(z) = 3 - 2z^{-1} + 2z^{-2} - z^{-3}.$$

The transform of the sequence of partial sums after modification to a rational function becomes

$$\mathscr{Z}\left\{\sum_{k=0}^{n-1} f_k\right\} = \frac{3z^3 - 2z^2 + 2z - 1}{z^3(z-1)}.$$

Now, we construct the sequence of partial sums $\{g_n\} = \left\{\sum_{k=0}^{n-1} f_k\right\}$ by successive substitution. We obtain $\{g_n\} = \{0, 3, 1, 3, 2, 2, \ldots,\}$. We see that in consequence of the fact that $f_n = 0$ for $n > 3$, we have $g_n = 2$ for $n \geqq 4$. Both sequences are indicated in Fig. 9.

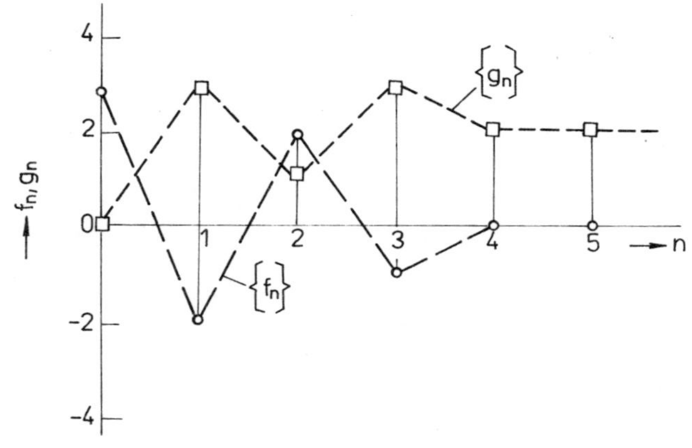

Fig. 9

EXAMPLE 2. In this example we use the $\mathscr{Z}$ transform to show the mutual inversity of the operations of difference and partial summation. Let us look for the transform of the sequence $\mathscr{Z}\left\{\Delta\sum_{k=0}^{n-1} f_k\right\}$. For the sequence of partial sums let us introduce the notation $\{g_n\}$. Then it is possible to write, applying the theorem on the transform of the difference,

$$\mathscr{Z}\{\Delta g_n\} = (z-1)\,G(z) - zg_0\,.$$

By Theorem 11 we obtain for $\mathscr{Z}\{g_n\}$

$$\mathscr{Z}\{g_n\} = \mathscr{Z}\left\{\sum_{k=0}^{n-1} f_k\right\} = \frac{F(z)}{z-1}\,.$$

The first term of the sequence $\{g_n\}$ follows from the summation and we have $g_0 = 0$. Upon substitution we arrive at the expected result

$$\mathscr{Z}\left\{\Delta \sum_{k=0}^{n-1} f_k\right\} = F(z) = \mathscr{Z}\{f_n\}.$$

EXAMPLE 3. Let us determine the transform of the sequence of sums $\sum_{k=0}^{n-1} k$. By relation (2.17) we have

$$\mathscr{Z}\{n\} = \frac{z}{(z-1)^2}$$

for the transform of the sequence $\{n\}$. The direct application of Theorem 11 yields the desired result

$$\mathscr{Z}\left\{\sum_{k=0}^{n-1} k\right\} = \frac{z}{(z-1)^3}. \tag{2.21}$$

If we compare this relation with the transform (2.18), it is obvious that as a consequence of the uniqueness of the transform the following equality holds:

$$\sum_{k=0}^{n-1} k = \binom{n}{2}.$$

2.1.8. *Differentiation of the Transform of a Sequence*

THEOREM 12. *If* $\mathscr{Z}\{f_n\} = F(z)$ *exists for* $|z| > 1/R$, *then* $\mathscr{Z}\{nf_n\}$ *also exists and for* $|z| > 1/R$ *we have the relation*

$$\mathscr{Z}\{nf_n\} = -z \frac{\mathrm{d}}{\mathrm{d}z} F(z). \tag{2.22}$$

EXAMPLE. Let us determine the transform of the sequence $\{n^2\}$. We already know the transform (2.17)

$$\mathscr{Z}\{n\} = \frac{z}{(z-1)^2}.$$

Direct application of Theorem 12 yields

$$\mathscr{Z}\{n^2\} = -z\frac{\mathrm{d}}{\mathrm{d}z}\frac{z}{(z-1)^2} = \frac{z(z+1)}{(z-1)^3}. \tag{2.23}$$

Repeated application results in

$$\mathscr{Z}\{n^3\} = -z\frac{\mathrm{d}}{\mathrm{d}z}\frac{z(z+1)}{(z-1)^3} = \frac{z(z^2+4z+1)}{(z-1)^4}. \tag{2.24}$$

The transform of the sequence $\{n^{i-1}\}$ can be written in the general form

$$\mathscr{Z}\{n^{i-1}\} = \frac{N_i(z)}{(z-1)^i}, \tag{2.25}$$

where $N_i(z)$ is a polynomial in the variable z. A table of its coefficients and zeros, for $i = 1, 2, \ldots, 10$, together with the corresponding recurrent relations is given in the Appendix.

2.1.9. *Integration of the Transform of a Sequence*

THEOREM 13. *Let the term $f_0 = 0$ be defined. If $\mathscr{Z}\{f_n\} = F(z)$ exists for $|z| > 1/R$, then $\mathscr{Z}(f_n/n)$ also exists and for $|z| > 1/R$ we have*

$$\mathscr{Z}\left\{\frac{f_n}{n}\right\} = \int_z^\infty \frac{F(\zeta)}{\zeta}\,\mathrm{d}\zeta. \tag{2.26}$$

Since the entire path of integration lies in the region of the regularity of the integrated function, the integral is independent of the integration path and it is a function of the endpoint z. In other words, we may integrate according to the same rules as in the case of a real variable.

EXAMPLE. The sequence $\{f_n\} = \{(-1)^{n-1}/n\}$ is given for $n \geqq 1$, $f_0 = 0$. With the aid of Theorem 13 we obtain its transform

$$\mathscr{Z}\left\{\frac{(-1)^{n-1}}{n}\right\} = \int_z^\infty \frac{\mathrm{d}\zeta}{\zeta(\zeta+1)} = -\ln\frac{z}{z+1} = \ln\left(1+\frac{1}{z}\right). \tag{2.27}$$

2.1.10. *Differentiation with Respect to a Parameter.*

THEOREM 14. *Let a sequence of functions $\{f_n(\xi)\}$ be given which are defined for all ξ from an interval $[a, b]$, and let the derivatives $f_n'(\xi)$ be finite* *in this interval. If* $\mathscr{Z}\{f_n(\xi)\} = F(z, \xi)$ *exists for at least one* ξ *from this interval, and if the series* $\sum_{n=0}^{+\infty} f_n'(\xi)\, z^{-n}$ *converges uniformly for* ξ *from the interval* $[a, b]$, *then the series* $\mathscr{Z}\{f_n(\xi)\}$ *also converges uniformly for* ξ *from the same interval and we have*

$$\frac{\mathrm{d}F(z,\xi)}{\mathrm{d}\xi} = \mathscr{Z}\left\{\frac{\mathrm{d}f_n(\xi)}{\mathrm{d}\xi}\right\}. \tag{2.28}$$

EXAMPLE. Let us look for the transform of the sequence of functions $\{(n+1)\,\xi^n\}$ for ξ which lie in the interval $[0, b]$ $(b < +\infty)$. Let us introduce a new sequence $\{f_n(\xi)\} = \{\xi^{n+1}\}$ for which we obviously have $f_n'(\xi) = (n+1)\,\xi^n$. Since the conditions of Theorem 14 are satisfied (this can be verified with the aid of Theorem 2) it is possible to exchange the order of differentiation and $\mathscr{Z}$ transformation and we obtain

$$\mathscr{Z}\{(n+1)\,\xi^n\} = \mathscr{Z}\left\{\frac{\mathrm{d}\xi^{n+1}}{\mathrm{d}\xi}\right\} = \frac{\mathrm{d}}{\mathrm{d}\xi}\,\mathscr{Z}\{\xi^{n+1}\} =$$

$$= \frac{\mathrm{d}}{\mathrm{d}\xi}\,\frac{z\xi}{z-\xi} = \frac{z^2}{(z-\xi)^2}.$$

2.1.11. *Integration with Respect to a Parameter*

THEOREM 15. *A sequence of continuous functions* $\{f_n(\xi)\}$ *is given which are defined for* ξ *from an interval* $[a, b]$. *Let* $\mathscr{Z}\{f_n(\xi)\} = F(z, \xi)$ *exist for* $|z| > 1/R$. *If the series* $\mathscr{Z}\{f_n(\xi)\}$ *converges uniformly for* ξ *from the interval* $[a, b]$, *then* $\mathscr{Z}\{\int_a^b f_n(\xi)\,\mathrm{d}\xi\}$ *also exists and we have*

$$\mathscr{Z}\left\{\int_a^b f_n(\xi)\,\mathrm{d}\xi\right\} = \int_a^b F(z,\xi)\,\mathrm{d}\xi\,. \tag{2.29}$$

EXAMPLE. We wish to determine the transform of the sequence $\{b^{n+1}/(n+1)\}$. For the application of Theorem 15 we introduce a new sequence of functions $f_n(\xi) = \xi^n$ for which

$$\int_0^b \xi^n\,\mathrm{d}\xi = \frac{b^{n+1}}{n+1}.$$

The sequence $\{f_n(\xi)\}$ is transformable for all $|z| > |\xi|$ and we have

$$\mathscr{Z}\{\xi^n\} = \frac{z}{z - \xi}.$$

It is possible to find such a number z_0 that for every ξ from the given interval the inequality

$$|\xi| < |z_0|$$

is satisfied. By Theorem 2, the series which defines the transform is uniformly convergent and it is possible to exchange the order of integration and transformation:

$$\mathscr{Z}\left\{\frac{b^{n+1}}{n+1}\right\} = \mathscr{Z}\left\{\int_0^b \xi^n \,\mathrm{d}\xi\right\} = \int_0^b \mathscr{Z}\{\xi^n\} \,\mathrm{d}\xi = \int_0^b \frac{z}{z-\xi}\,\mathrm{d}\xi =$$

$$= [-z \ln(z - \xi)]_0^b = z \ln \frac{z}{z-b}. \qquad (2.30)$$

2.1.12. *Limit of the Sequence of Partial Sums*

THEOREM 16. *If the series* $\sum_{n=0}^{+\infty} f_n$ *converges and if* $\mathscr{Z}\{f_n\} = F(z)$, *then we have the equality*

$$\lim_{\substack{z \to 1+ \\ \mathrm{Im}\, z = 0}} F(z) = \sum_{n=0}^{+\infty} f_n. \qquad (2.31)$$

This theorem is exploited with advantage for the computation of the sum of an infinite series if the transform of the sequence of its coefficients is known.

EXAMPLE 1. We look for the sum of the series $\sum_{n=1}^{+\infty} (-1)^{n-1}/n$ which is convergent but not absolutely convergent. Using the result (2.27) of the example from Subsection 2.1.9

$$\mathscr{Z}\left\{\frac{(-1)^{n-1}}{n}\right\} = \ln\left(1 + \frac{1}{z}\right)$$

we obtain, passing to the limit for $z \to 1+$ along the real axis,

$$\sum_{n=1}^{+\infty} \frac{(-1)^{n-1}}{n} = \lim_{\substack{z \to 1+ \\ \operatorname{Im} z = 0}} \sum_{n=1}^{+\infty} \frac{(-1)^{n-1}}{n} z^{-n} = \lim_{\substack{z \to 1+ \\ \operatorname{Im} z = 0}} \ln\left(1 + \frac{1}{z}\right) = \ln 2 .$$

EXAMPLE 2. Similarly, we easily determine the sum of the series $\sum_{n=0}^{+\infty} x^{n+1}/(n+1)$ for $-1 \leqq x < 1$. For this we use the transform of the sequence (2.30) which was obtained in the example of Subsection 2.1.11:

$$\sum_{n=0}^{+\infty} \frac{x^{n+1}}{n+1} = \lim_{\substack{z \to 1+ \\ \operatorname{Im} z = 0}} z \ln \frac{z}{z - x} = \ln \frac{1}{1 - x} .$$

2.1.13. *Limit Values of the Object Function*

THEOREM 17. *If the transform* $\mathscr{Z}\{f_n\} = F(z)$ *exists for* $|z| > 1/R$, *then*

$$\lim_{z \to \infty} F(z) = f_0 . \tag{2.32}$$

THEOREM 18. *Let* $\mathscr{Z}\{f_n\} = F(z)$ *for* $|z| > 1/R$. *If* $\lim_{n \to +\infty} f_n$ *exists, then* $\lim_{\substack{z \to 1+ \\ \operatorname{Im} z = 0}} (z - 1) F(z)$ *also exists and we have*

$$\lim_{\substack{z \to 1+ \\ \operatorname{Im} z = 0}} (z - 1) F(z) = \lim_{n \to +\infty} f_n . \tag{2.33}$$

EXAMPLE. We wish to determine the limit values of the transform of the sequence of partial sums constructed in Example 1 of Subsection 2.1.7:

$$G(z) = \frac{3z^3 - 2z^2 + 2z - 1}{z^3(z - 1)} .$$

Since the degree of the numerator is smaller than the degree of the denominator, it is obvious that $\lim_{z \to \infty} G(z) = g_0 = 0$. Since we know that the sum of a finite sequence exists, $\lim_{n \to +\infty} g_n$ also exists and we have

$$\lim_{n \to +\infty} g_n = \lim_{\substack{z \to 1+ \\ \operatorname{Im} z = 0}} (z - 1) \frac{3z^3 - 2z^2 + 2z - 1}{z^3(z - 1)} = 2 .$$

2.2. Inverse $\mathscr{Z}$ Transform

By the *inverse $\mathscr{Z}$ transform* we understand the procedure of the determination of the object function, i.e. of the sequence $\{f_n\}$, to the given transform $F(z)$. We use the symbolic notation

$$\{f_n\} = \mathscr{Z}^{-1}\{F(z)\}\,.$$

The inverse $\mathscr{Z}$ transform can be obtained in different ways; all make use of the uniqueness of the $\mathscr{Z}$ transform (Theorems 4a, 4b). The simplest approach is to use the list of transforms. However, if the transform is not located in the list we try to decompose it into simpler partial transforms which are included in the list. If the transform is decomposed into a product of partial sums, the resulting object function is obtained as the convolution of the partial object functions (see Example 2 in Subsection 2.1.5). On the other hand, if the decomposition is a sum of partial transforms, we obtain the desired object function by the summation of the partial object functions.

A typical example of the decomposition into a linear combination of simpler transforms is the decomposition of a rational function into partial fractions. This we demonstrate on simple examples.

EXAMPLE 1. We are given the transform

$$F(z) = \frac{z-1}{(z+1)(z-0.5)}$$

and we look for the sequence $\{f_n\} = \mathscr{Z}^{-1}\{F(z)\}$. We decompose the transform into two partial functions

$$\frac{z-1}{(z+1)(z-0.5)} = \frac{A}{z+1} + \frac{B}{z-0.5}\,.$$

Comparing the coefficients at the same powers of the variable z we obtain two linear equations for the unknown coefficients A and B:

$$\begin{aligned} z-1 &= A(z-0.5) + B(z+1)\,, \\ 1 &= A + B\,, \\ -1 &= -0.5A + B\,. \end{aligned}$$

The solution of these equations yields $A = 4/3$, $B = -1/3$. Using the list of transforms and the translation theorem we obtain

$$\mathscr{Z}^{-1}\left\{\frac{4}{3}\frac{1}{z+1}\right\} = \frac{4}{3}(-1)^{n-1},$$

$$\mathscr{Z}^{-1}\left\{-\frac{1}{3}\frac{1}{z-0.5}\right\} = -\frac{1}{3}0.5^{n-1}$$

for $n \geqq 1$. The initial value f_0 is easily determined from the original transform applying Theorem 17:

$$f_0 = \lim_{z \to \infty} F(z) = 0.$$

As far as a transform in the form of a rational function has a zero at the origin, i.e. the numerator contains the coefficient z, it is advantageous to factor this zero out when performing the decomposition. Let us show this on a modification of the above example.

EXAMPLE 2. We are given the transform

$$G(z) = \frac{z(z-1)}{(z+1)(z-0.5)}.$$

The sequence $\{g_n\}$ is obtained by the inverse transform of the decomposition

$$\mathscr{Z}\{G(z)\} = \mathscr{Z}\{z\,F(z)\} = \mathscr{Z}\left\{\frac{Az}{z+1} + \frac{Bz}{z-0.5}\right\},$$

where $F(z)$ is the transform from Example 1. With the aid of the list of transforms we obtain

$$\{g_n\} = \mathscr{Z}^{-1}\left\{\frac{4}{3}\frac{z}{z+1}\right\} + \mathscr{Z}^{-1}\left\{-\frac{1}{3}\frac{z}{z-0.5}\right\} =$$

$$= \frac{4}{3}(-1)^n - \frac{1}{3}0.5^n \qquad \text{for} \qquad n \geqq 0.$$

For the initial term of the sequence we now have $g_0 = 1$; this can be also verified with the aid of Theorem 17.

2.2.1. *Inverse Transform with the Aid of Integration*

The general approach to the inverse transform consists in the application of the Cauchy integral theorem to the determination of the coefficients of the Laurent expansion. In the case of the one-sided $\mathscr{Z}$ transform the following theorem holds:

THEOREM 19. *Let $F(z)$ be a regular function in the region $|z| > 1/R$. Then there exists a single sequence $\{f_n\}$ for which $\mathscr{Z}\{f_n\} = F(z)$, namely*

$$f_n = \frac{1}{2\pi j} \oint_C F(z)\, z^{n-1}\, dz \qquad \textit{for} \qquad n = 0, 1, 2, \ldots, \tag{2.34}$$

$$f_n = 0 \qquad \textit{for} \qquad n < 0 .$$

Integration is performed along the circle C determined by $z = \varrho\, e^{j\varphi}$, where $\varrho > 1/R$ and $0 \leqq \varphi \leqq 2\pi$.

In thė case of a rational function $F(z)$ the residue theorem is applied with advantage to the solution of the integral. Without loss of generality, we may assume that the transform $F(z)$ has the form

$$F(z) = \frac{P(z)}{Q(z)} = \frac{a_0 z^s + a_1 z^{s-1} + \ldots + a_{s-1} z + a_s}{z^s + b_1 z^{s-1} + \ldots + b_{s-1} z + b_s},$$

where the degrees of the polynomials in the numerator and in the denominator were chosen the same, and the coefficient at z^s in the denominator was chosen equal to one.

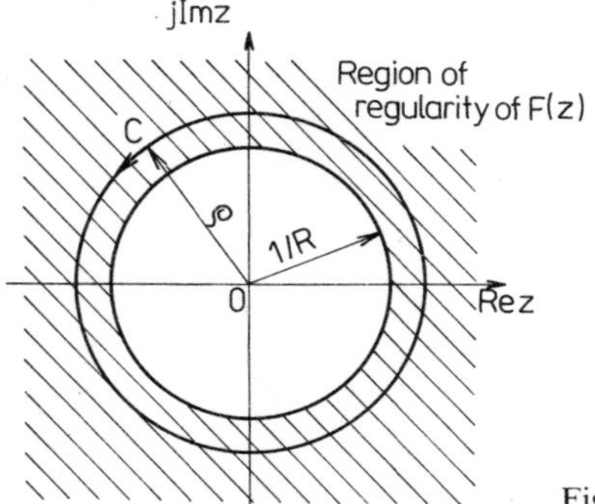

Fig. 10

The poles of the function $F(z)$ are given by the roots of the equation $Q(z) = 0$. If the coefficients b_i, $i = 1, 2, \ldots, s$, are real numbers, the poles are either real or they occur in complex conjugate pairs. If the coefficients a_i are real numbers, the same is true for the zeros of the function $F(z)$.

If we apply the residue theorem to the integral of the inverse transform, we obtain

$$f_n = \frac{1}{2\pi j} \oint_C F(z)\, z^{n-1}\, dz = \sum_{i=1}^{K} \operatorname*{res}_{z=z_i} F(z)\, z^{n-1}, \tag{2.35}$$

where K is the number of different poles z_i of the function $F(z)\, z^{n-1}$ which lie in the circle with radius $\varrho > 1/R$ (see Fig. 10).

For the residue at the pole z_i with multiplicity m of the function $F(z)\, z^{n-1}$ we have the relation

$$\operatorname*{res}_{z=z_i} F(z)\, z^{n-1} = \\ = \frac{1}{(m-1)!} \lim_{z \to z_i} \frac{d^{m-1}}{dz^{m-1}} \left[(z - z_i)^m F(z)\, z^{n-1}\right], \tag{2.36}$$

which may be modified, applying the Leibnitz rule for the computation of the derivative of order $m - 1$, to the form

$$\operatorname*{res}_{z=z_i} F(z)\, z^{n-1} = \\ = \sum_{r=1}^{m} \frac{1}{(m-r)!} \binom{n-1}{r-1} z^{n-r} \left[\frac{d^{m-r}}{dz^{m-r}} (z - z_i)^m F(z)\right]\bigg|_{z=z_i}. \tag{2.37}$$

For a simple pole, i.e. for $m = 1$, relation (2.36) reduces to

$$\operatorname*{res}_{z=z_i} F(z)\, z^{n-1} = \lim_{z \to z_i} (z - z_i)\, F(z)\, z^{n-1} = \frac{P(z_i)}{Q'(z_i)}\, z_i^{n-1}, \tag{2.38}$$

where

$$Q'(z_i) = \frac{dQ(z)}{dz}\bigg|_{z=z_i}.$$

The expression on the right-hand side of relation (2.38) is preferable if the function $Q(z)$ has the form of a polynomial; on the other hand, the relation with the limit is used if the function $Q(z)$ is given as a product of the root factors since then it is possible to cancel out the factor $z - z_i$.

When performing the inverse transform as based on Theorem 19 we have to realize that the existence and multiplicity of a pole at the origin of the integrated function $F(z)\,z^{n-1}$ depends on the subscript n. Namely, if the transform $F(z)$ does not have a zero at $z = 0$ but has a pole of multiplicity m there, it is more suitable to compute the initial values of the sequence $\{f_n\}$ individually. Only for $n - 1 - m \geqq 0$, i.e. for $n \geqq m + 1$, relation (2.35) represents a formula for the terms of the sequence $\{f_n\}$. Sometimes it is possible to represent the transform $F(z)$ by a product $F(z) = z^{-m}\,G(z)$ where $G(z)$ is a regular function for $|z| > 1/R$, thus it is the transform of the sequence $\{g_n\}$. Finding the inverse $\mathscr{Z}$ transform to the transform $G(z) = z^m\,F(z)$ and applying the translation theorem, we obtain

$$\{f_n\} = \{g_{n-m}\}\,.$$

Another approach to the determination of the initial m terms of the sequence $\{f_n\}$ consists in the application of the inverse transform procedure according to Subsection 2.2.2.

Thus, relation (2.35) renders a formula for the terms f_n for all $n \geqq 0$ only in the case that the transform $F(z)$ has at least a simple zero at $z = 0$.

The application of Theorem 19 will be demonstrated on examples from Section 2.2.

EXAMPLE 1. We are given the transform

$$F(z) = \frac{z(z - 1)}{(z + 1)(z - 0.5)}$$

which is a regular function for $|z| > 1$. With the aid of the residue theorem and relation (2.38) we obtain

$$f_n = \operatorname*{res}_{z=-1} \frac{(z - 1)}{(z + 1)(z - 0.5)}\,z^n + \operatorname*{res}_{z=0.5} \frac{(z - 1)}{(z + 1)(z - 0.5)}\,z^n =$$

$$= \frac{4}{3}(-1)^n - \frac{1}{3}\,0.5^n\,, \qquad n = 0, 1, 2, \ldots.$$

EXAMPLE 2. If we omit the zero at the origin in the previous example, we obtain the transform

$$F(z) = \frac{z - 1}{(z + 1)(z - 0.5)}\,.$$

For the object function we now have

$$f_0 = \operatorname*{res}_{z=0} \frac{z-1}{(z+1)(z-0.5)\,z} + \operatorname*{res}_{z=-1} \frac{z-1}{(z+1)(z-0.5)\,z} + $$
$$+ \operatorname*{res}_{z=0.5} \frac{z-1}{(z+1)(z-0.5)\,z} = 0\,,$$

and for $n \geqq 1$

$$f_n = \operatorname*{res}_{z=-1} \frac{z-1}{(z+1)(z-0.5)} z^{n-1} + \operatorname*{res}_{z=0.5} \frac{z-1}{(z+1)(z-0.5)} z^{n-1} =$$
$$= \frac{4}{3}(-1)^{n-1} - \frac{1}{3} 0.5^{n-1}\,.$$

EXAMPLE 3. As an example of the inverse transform with a pole at the origin we present the transform

$$F(z) = \frac{(z-1)^3}{z(z+1)(z-0.5)}\,.$$

The degree of the numerator is equal to the degree of the denominator. Therefore we do not employ the translation theorem but directly relations (2.35) and (2.36). We obtain

$$f_0 = \operatorname*{res}_{z=0} \frac{(z-1)^3}{z^2(z+1)(z-0.5)} + \operatorname*{res}_{z=-1} \frac{(z-1)^3}{z^2(z+1)(z-0.5)} +$$
$$+ \operatorname*{res}_{z=0.5} \frac{(z-1)^3}{z^2(z+1)(z-0.5)} = -4 + \frac{16}{3} - \frac{1}{3} = 1\,,$$

$$f_1 = \operatorname*{res}_{z=0} \frac{(z-1)^3}{z(z+1)(z-0.5)} + \operatorname*{res}_{z=-1} \frac{(z-1)^3}{z(z+1)(z-0.5)} +$$
$$+ \operatorname*{res}_{z=0.5} \frac{(z-1)^3}{z(z+1)(z-0.5)} = 2 - \frac{16}{3} - \frac{1}{6} = -3.5\,,$$

and for $n \geqq 2$

$$f_n = \operatorname*{res}_{z=-1} \frac{(z-1)^3}{(z+1)(z-0.5)} z^{n-2} + \operatorname*{res}_{z=0.5} \frac{(z-1)^3}{(z+1)(z-0.5)} =$$

$$= \frac{16}{3}(-1)^{n-2} - \frac{1}{12} 0.5^{n-2} .$$

2.2.2. *Inverse Transform with the Aid of Differentiation*

The coefficients of the Laurent expansion of a function $F(z)$ which is regular for $|z| > 1/R$ may be also determined with the aid of differentiation. For the one-sided $\mathscr{Z}$ transform we have

THEOREM 20. *Let $F(z)$ be a regular function in the region $|z| > 1/R$. Then there exists a unique sequence $\{f_n\}$ $(n \geqq 0)$ for which $\mathscr{Z}\{f_n\} = F(z)$, namely*

$$f_n = \frac{1}{n!} \frac{\mathrm{d}^n}{\mathrm{d}\zeta^n} F\left(\frac{1}{\zeta}\right)\bigg|_{\zeta=0} \qquad \textit{for} \qquad n = 0, 1, 2, \ldots, \tag{2.39}$$

$$f_n = 0 \qquad \textit{for} \qquad n < 0 .$$

EXAMPLE. The direct application of Theorem 20 will be demonstrated using the transform given by the transcendental function

$$F(z) = \sin \frac{1}{z},$$

which is regular for all $|z| > 0$. Computing several derivatives we find out that

$$F(z) = z^{-1} - \frac{1}{3!} z^{-3} + \frac{1}{5!} z^{-5} - \ldots .$$

Thus, the desired sequence is given by the formula

$$f_m = 0 \qquad \text{for} \qquad m = 2n ,$$

$$f_m = \frac{(-1)^{(m-1)/2}}{m!} \qquad \text{for} \qquad m = 2n + 1, \quad n = 0, 1, 2, \ldots .$$

The application of Theorem 20 to a transform which has the form of a rational function does not require the knowledge of the poles. Thus, the solution of the equation $Q(z) = 0$ is not necessary. However, the computation of higher derivatives of the function $F(1/\zeta)$ is complicated and tedious. For this reason it is advantageous to use some other – algebraic – method of the inverse transform which would help us determine all the terms of the sequence $\{f_n\}$ by a fast recurrent procedure based directly on the coefficients of the rational function, without computing the derivatives. This approach will be called the *numerical method of the inverse $\mathscr{Z}$ transform.*

Consider again the transform $F(z)$ in the form

$$F(z) = \frac{P(z)}{Q(z)} = \frac{a_0 z^s + a_1 z^{s-1} + \dots + a_{s-1} z + a_s}{z^s + b_1 z^{s-1} + \dots + b_{s-1} z + b_s}.$$

If we divide the numerator by the denominator starting with the highest powers of the variable z, we obtain a series in the powers of z^{-1}. In consequence of the uniqueness of the $\mathscr{Z}$ transform its coefficients are identical with the terms of the sequence $\{f_n\}$, where the subscript of the term of the sequence is equal to the exponent of the variable z^{-1}. The elementary approach to the division of a polynomial $P(z)$ by a polynomial $Q(z)$ has the rather complicated notation

$$\begin{array}{llll}
(a_0 z^s + & a_1 z^{s-1} + \dots + & a_s) : (z^s + b_1 z^{s-1} + \dots + b_s) = & \\
\underline{\pm a_0 z^s \pm} & \underline{a_0 b_1 z^{s-1} + \dots \pm} & \underline{a_0 b_s} & = a_0 + (a_1 - a_0 b_1) z^{-1} + \dots . \\
 & (a_1 - a_0 b_1) z^{s-1} + \dots + & (a_s - a_0 b_s) & \\
 & \vdots & \vdots &
\end{array}$$

Comparing the ratio with the definition of the $\mathscr{Z}$ transform we obtain $f_0 = a_0$, $f_1 = (a_1 - a_0 b_1)$, ... The evaluation corresponds to the substitution into the recurrent relation

$$f_n = a_n - \sum_{i=1}^{s} b_i f_{n-i}, \tag{2.40}$$

where $a_n = a_i$ for $n = i = 0, 1, 2, \dots, s$, while $a_n = 0$ for $n > s$. The computations can be mechanized with the aid of the so-called *strip method.*

Numerator	$\{f_n\}$
a_0	$f_0 = a_0$
a_1	$f_1 = a_1 - b_1 f_0$
a_2	$f_2 = a_2 - b_1 f_1 - b_2 f_0$
$\vdots$	$\vdots$
a_s	$f_s = a_s - \sum_{i=1}^{s} b_i f_{s-i}$
	$\vdots$
	$f_n = \sum_{i=1}^{s} b_i f_{n-i}$

Denominator
$-b_s$
$\vdots$
$-b_2$
$-b_1$
←

Strip
of paper

EXAMPLE 1. We demonstrate the application of the procedure on Example 2 of Subsection 2.2.1. where we had

$$F(z) = \frac{z-1}{(z+1)(z-0.5)} = \frac{z-1}{z^2 + 0.5z - 0.5}.$$

The corresponding recurrent relation is

$$f_n = a_n - 0.5f_{n-1} + 0.5f_{n-2}$$

for $a_0 = 0$, $a_1 = 1$, $a_2 = -1$, and $a_n = 0$ for $n \geqq 3$. The calculations are performed by the strip method

Numerator	$\{f_n\}$	
0	0	$= f_0$
1	1	$= f_1$
-1	-1.5	$= f_2$
	1.25	$= f_3$
	-1.375	$= f_4$
	1.312 5	$= f_5$
	$-1.343\ 75$	$= f_6$
	$\vdots$	

Denominator
0.5
-0.5
←

When the inverse transform of a rational function is carried out recurrently, only selected terms of the sequence $\{f_n\}$ interest us sometimes, e.g., only terms with even subscript, i.e. f_n for $n = 2k$, $k = 0, 1, 2, \ldots$. Thus, we wish to determine the sequence $\{h_n\}$ given by the formula

$$\begin{aligned} h_n &= f_n \qquad \text{for} \qquad n = 2k\,, \\ h_n &= 0 \qquad \text{for} \qquad n = 2k + 1, \quad k = 0, 1, 2, \ldots\,, \end{aligned} \tag{2.41a}$$

which can be expressed as

$$h_n = \frac{1}{2}\left[f_n + (-1)^n f_n\right], \qquad n = 0, 1, 2, \ldots\,. \tag{2.41b}$$

By the $\mathscr{Z}$ transform and the application of the similarity theorem we obtain the transform of the sequence $\{h_n\}$ in the form

$$\mathscr{Z}\{h_n\} = H(z) = \frac{1}{2}\left[F(z) + F(-z)\right]. \tag{2.42}$$

The function $H(z)$ includes only even powers of the variable z. The recurrent procedure of the inverse transform yields the sequence (2.41). It is possible to repeat the procedure with the function $H(z)$; we obtain the transform the inverse of which yields only the terms of the original sequence $\{f_n\}$ for $n = 4k$, $k = 0, 1, 2, \ldots$.

EXAMPLE 2. The computation of the terms of the decimated sequence is shown using the preceding example of the transform

$$F(z) = \frac{z - 1}{(z + 1)(z - 0.5)}.$$

The transform $H(z)$ is

$$\begin{aligned} H(z) &= \frac{1}{2}\left[F(z) + F(-z)\right] = \\ &= \frac{1}{2}\left[\frac{z - 1}{(z + 1)(z - 0.5)} + \frac{-z - 1}{(-z + 1)(-z - 0.5)}\right] = \\ &= \frac{-1.5z^2 + 0.5}{z^4 - 1.25z^2 + 0.25}. \end{aligned}$$

The series expansion is executed using the strip method:

Numerator	$\{h_n\}$	
0	0	$= h_0 = f_0$
0	0	$= h_1$
-1.5	-1.5	$= h_2 = f_2$
0	0	$= h_3$
0.5	-1.375	$= h_4 = f_4$
	0	$= h_5$
	$-1.343\,75$	$= h_6 = f_6$

Denominator
-0.25
0
1.25
0
←

It is obvious that the number of arithmetic operations carried out when expanding $H(z)$ is the same as for the expansion of $F(z)$.

Now, let us discuss the algebraic inverse transform of the product of two transforms $F(z) = \mathscr{Z}\{f_n\}$, $G(z) = \mathscr{Z}\{g_n\}$. According to Theorem 9 it holds that the product of the transforms is the transform of the convolution

$$H(z) = F(z)\,G(z) = \mathscr{Z}\left\{\sum_{k=0}^{n} f_k g_{n-k}\right\} = \mathscr{Z}\{h_n\}\,.$$

On the whole, three cases may arise:

(a) Both transforms are in the form of a series in the variable z^{-1}, i.e.

$$F(z) = \sum_{n=0}^{+\infty} f_n z^{-n} \qquad \text{and} \qquad G(z) = \sum_{n=0}^{+\infty} g_n z^{-n}\,.$$

The product of the transforms is equal to the product of the two series, for the nth term of the sequence $\{h_n\}$ we have the convolution sum

$$h_n = \sum_{k=0}^{n} f_k g_{n-k}\,.$$

The evaluation of the terms of the sequence $\{h_n\}$ we perform efficiently with the aid of the modified strip method.

$\{g_n\}$	$\{f_n\}$	$\{h_n\}$
g_n	f_0	$h_0 = g_0 f_0$
g_{n-1}	f_1	$h_1 = g_0 f_1 + g_1 f_0$
$\vdots$	f_2	$h_2 = g_0 f_2 + g_1 f_1 + g_2 f_0$
	$\vdots$	$\vdots$
g_1	f_{n-1}	
$g_0 \rightarrow$	f_n	$h_n = \sum_{k=0}^{n} g_k f_{n-k}$
	$\vdots$	

Strip
of paper

(b) The transform $F(z)$ is in the form of a series in the variable z^{-1}, while the transform $G(z)$ is in the form of a rational function in the variable z.

The computation of the terms of the sequence consists in the multiplication of the series $F(z)$ by the polynomial in the numerator of the transform $G(z)$, and the division of the result by the polynomial in the denominator of the transform $G(z)$. It is also possible to expand the transform $G(z)$ into a series first (see the numerical inverse transform), and then to multiply it by the series for $F(z)$ [see the preceding case of (a)]. However, this procedure leads to a larger number of arithmetic operations.

Assume the transform $G(z)$ in the form

$$G(z) = \frac{a_0 z^s + a_1 z^{s-1} + \ldots + a_{s-1} z + a_s}{z^s + b_1 z^{s-1} + \ldots + b_{s-1} z + b_s}$$

and $F(z) = \sum_{n=0}^{+\infty} f_n z^{-n}$. The computation is organized as follows:

Paper template	$\{f_n\}$	$\{h_n\}$	
Numerator	f_0	$h_0 = a_0 f_0$	Denominator
	f_1	$h_1 = a_0 f_1 + a_1 f_0 - b_1 h_0$	
	f_2	$h_2 = a_0 f_2 + a_1 f_1 + a_2 f_0 - b_1 h_1 - b_2 h_0$	
a_s	$\vdots$	$\vdots$	$-b_s$
$\vdots$			$\vdots$
a_2	f_{n-2}	h_{n-2}	$-b_2$
a_1	f_{n-1}	h_{n-1}	$-b_1$
$a_0 \rightarrow$		$h_n = \sum_{i=0}^{s} a_i f_{n-i} - \sum_{i=1}^{s} b_i h_{n-i}$	$\leftarrow$

(c) Both transforms are rational functions of the variable z.

The computation of the sequence $\{h_n\}$ consists in the inverse $\mathscr{Z}$ transform of the transform given by the product of $F(z)$ and $G(z)$. This can be achieved either in accordance with the theorem on the inverse $\mathscr{Z}$ transform (we obtain a formula for the sequence $\{h_n\}$), or recurrently with the aid of the algorithm for the numerical inverse transform, or by the expansion of both or only one of the transforms into a series followed by the application or either algorithm (a) or algorithm (b).

If any of the transforms, e.g. $G(z)$, is given in the form of a product of simpler transforms $G(z) = \prod_{i=1}^{r} G_i(z)$, for the transform of the convolution we have

$$H(z) = F(z)\,G(z) = F(z)\prod_{i=1}^{r} G_i(z)\,.$$

This means that we have to perform r convolutions by the formulae

$$\begin{aligned}
H_1(z) &= F(z)\,G_1(z)\,,\\
H_2(z) &= H_1(z)\,G_2(z)\,,\\
&\vdots\\
H_{r-1}(z) &= H_{r-2}(z)\,G_{r-1}(z)\,,\\
H(z) = H_r(z) &= H_{r-1}(z)\,G_r(z)\,.
\end{aligned}$$

In the frequent case when the transform $F(z)$ is in the form of a series while the transforms $G_i(z)$ are rational functions, we compute the convolution by the repeated application of procedure (b).

2.3. Transform of the Product of Two Sequences

THEOREM 21. *If there exist* $\mathscr{Z}\{f_n\} = F(z)$ *for* $|z| > 1/R_1 > 0$ *and* $\mathscr{Z}\{g_n\} = G(z)$ *for* $|z| > 1/R_2 > 0$, *then* $\mathscr{Z}\{f_n g_n\}$ *also exists for* $|z| > 1/(R_1 R_2)$ *and the relation holds*

$$\mathscr{Z}\{f_n g_n\} = \frac{1}{2\pi \mathrm{j}} \oint_C \frac{F(\zeta)\, G(z/\zeta)}{\zeta} \,\mathrm{d}\zeta\,. \tag{2.43}$$

Integration is performed along the circle C given by $\zeta = \varrho\, \mathrm{e}^{\mathrm{j}\varphi}$, $1/R_1 < \varrho < |z|\, R_2$, $0 \leqq \varphi \leqq 2\pi$.

In other words: Integration is performed in the positive sense along the circle inside which lie all the singular points of the function $F(\zeta)$ and outside which lie all the singular points of the function $G(z/\zeta)$ (see Fig. 11). Theorem 21 is also referred to as the *theorem on the convolution of transforms.*

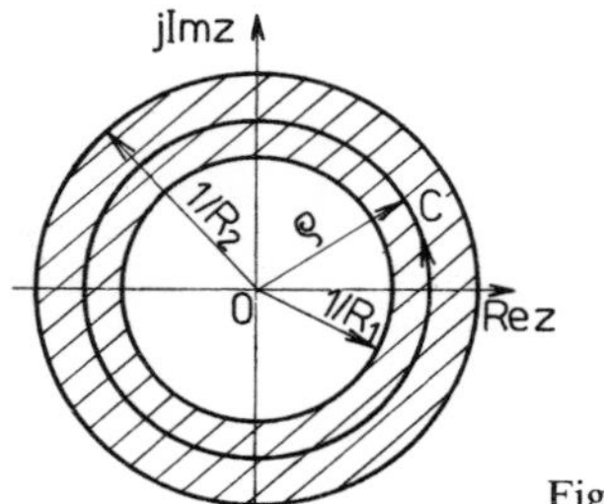

Fig. 11

As a rule, the integral is solved with the aid of the residue theorem which yields in this case

$$\mathscr{Z}\{g_n f_n\} = \sum_{i=1}^{K} \operatorname*{res}_{\zeta=\zeta_i} \frac{F(\zeta)\, G(z/\zeta)}{\zeta}, \tag{2.44}$$

where K is the number of different poles ζ_i $(i = 1, 2, \ldots, K)$ of the function $F(\zeta)/\zeta$. For the residue at the pole ζ_i of multiplicity m of the function $F(\zeta)/\zeta$ we have

$$\operatorname*{res}_{\zeta=\zeta_i} \frac{F(\zeta)\, G\left(\frac{z}{\zeta}\right)}{\zeta} = \frac{1}{(m-1)!} \lim_{\zeta \to \zeta_i} \frac{\mathrm{d}^{m-1}}{\mathrm{d}\zeta^{m-1}} \left[(\zeta - \zeta_i)^m \frac{F(\zeta)\, G\left(\frac{z}{\zeta}\right)}{\zeta} \right], \tag{2.45}$$

Hence, for a simple pole for $m = 1$ we obtain

$$\operatorname*{res}_{\zeta=\zeta_i} \frac{F(\zeta)\, G\left(\frac{z}{\zeta}\right)}{\zeta} = \lim_{\zeta \to \zeta_i} (\zeta - \zeta_i) \frac{F(\zeta)\, G\left(\frac{z}{\zeta}\right)}{\zeta} = G\left(\frac{z}{\zeta}\right) \lim_{\zeta \to \zeta_i} (\zeta - \zeta_i) \frac{F(\zeta)}{\zeta}. \tag{2.46}$$

Theorem 21 is used, as a rule, for the determination of transforms of more complicated sequences.

EXAMPLE 1. Let us determine the transform of the sequence $\{n\, \mathrm{e}^{an}\}$. From the preceding examples we know that

$$\mathscr{Z}\{n\} = \frac{z}{(z-1)^2} \quad \text{for} \quad |z| > 1\,,$$

$$\mathscr{Z}\{\mathrm{e}^{an}\} = \frac{z}{z - \mathrm{e}^a} \quad \text{for} \quad |z| > |\mathrm{e}^a|\,.$$

We put $\mathscr{Z}\{\mathrm{e}^{an}\} = F(z)$ and $\mathscr{Z}\{n\} = G(z)$. Thus, we have chosen as the function $F(z)$ a transform with a simple pole which enables the application of the simpler relation for the evaluation of the residue when computing the integral. We obtain

$$\mathscr{Z}\{n\, \mathrm{e}^{an}\} = \frac{1}{2\pi \mathrm{j}} \oint_C \frac{z}{\zeta(\zeta - \mathrm{e}^a)(z/\zeta - 1)^2} \,\mathrm{d}\zeta\,.$$

For the radius of the integration circle the relation $|e^a| < \varrho < |z|$ must hold. The residue theorem and relation (2.46) imply

$$\mathscr{Z}\{n\,e^{an}\} = \operatorname*{res}_{\zeta=e^{\alpha}} \frac{z\zeta}{(\zeta - e^a)(z-\zeta)^2} = \frac{z\,e^a}{(z-e^a)^2}\,. \tag{2.47}$$

The same result would be obtained by the application of Theorem 8 or of Theorems 12 or 14.

With the aid of Theorem 21 it is possible to construct a decimated sequence $\{h_n\}$ to a given sequence $\{f_n\}$ in a general manner. In the preceding examples this problem was encountered already a number of times.

EXAMPLE 2. Let the transform of a sequence $\mathscr{Z}\{f_n\} = F(z)$ be given for $|z| > 1/R_1$. For a sequence $\{g_n\}$ selected from the unit sequence we have

$$\begin{aligned} g_n &= 1 \qquad \text{for} \qquad n = Nk\,, \\ g_n &= 0 \qquad \text{for} \qquad n \neq Nk, \quad k = 0, 1, 2, \ldots, \end{aligned}$$

i.e. $\{g_n\} = \{1, \underbrace{0, 0, \ldots, 0}_{(N-1)\text{ times}}, 1, \underbrace{0, 0, \ldots, 0}_{(N-1)\text{ times}}, 1, \ldots\}$.

By Theorem 7 we have

$$\mathscr{Z}\{g_n\} = G(z) = \frac{z^N}{z^N - 1} \qquad \text{for} \qquad |z| > 1\,.$$

The function $G(z)$ has only simple poles at the points of the roots of the equation $z^N - 1 = 0$, i.e. at the points $z_i = e^{j \cdot 2\pi i/N}$, $i = 0, 1, \ldots, N-1$. Obviously, the selected sequence from the sequence $\{f_n\}$ will be obtained by the product $\{f_n g_n\}$. The transform of the selected sequence is determined with the aid of Theorem 21:

$$\begin{aligned} \mathscr{Z}\{h_n\} &= \frac{1}{2\pi j} \oint_C F\left(\frac{z}{\zeta}\right) \frac{\zeta^{N-1}}{\zeta^N - 1}\,d\zeta = \sum_{i=0}^{N-1} \operatorname*{res}_{\zeta=\zeta_i} F\left(\frac{z}{\zeta}\right) \frac{\zeta^{N-1}}{\zeta^N - 1} = \\ &= \sum_{i=0}^{N-1} F\left(\frac{z}{\zeta_i}\right) \frac{\zeta_i^{N-1}}{\zeta_i^{N-1} + \zeta_i^{N-2} + \ldots + 1} = \\ &= \sum_{i=0}^{N-1} F\left(\frac{z}{\zeta_i}\right) \frac{\zeta_i^{N-1}}{\left.\dfrac{d(\zeta^N - 1)}{d\zeta}\right|_{\zeta=\zeta_i}} = \frac{1}{N} \sum_{i=0}^{N-1} F\left(\frac{z}{\zeta_i}\right). \end{aligned} \tag{2.48}$$

For $N = 2$, the poles of the function $G(z)$ lie at the points $z_{1,2} = \pm 1$ and the transform of the decimated sequence is

$$\mathscr{Z}\{h_n\} = \frac{1}{2}[F(z) + F(-z)],$$

This relation is identical with the relation obtained in Subsection 2.2.2.

2.3.1. *Transform of a Correlation Sequence*

As a general example of the application of the theorem on the convolution of transforms we have the representation of the correlation sequence $\{\psi_{fg,n}\}$ based on the transforms of the sequences $\{f_n\}$ and $\{g_n\}$.

An *aperiodic correlation sequence* is defined by the infinite sum

$$\psi_{fg,n} = \sum_{k=0}^{+\infty} f_k g_{k-n}. \tag{2.49}$$

For a sequence defined in this way we have

$$\psi_{fg,n} = \sum_{k=0}^{+\infty} f_k g_{k+n} = \sum_{k=0}^{+\infty} f_{k-n} g_k = \psi_{gf,n}. \tag{2.50}$$

Besides the exchange of the subscripts we employ here the fundamental assumption of the one-sidedness of the sequence, i.e. $f_n = 0$ and $g_n = 0$ for $n < 0$. Thus, it is possible to determine the sequence $\{\psi_{fg,n}\}$ for negative subscripts with the aid of the sequence $\{\psi_{gf,n}\}$ for positive subscripts. As a consequence, we may limit our interest to one of the two branches only.

Assume that to the two sequences there exist the transforms $\mathscr{Z}\{f_n\} = F(z)$ and $\mathscr{Z}\{g_n\} = G(z)$ for $|z| \geqq 1$. Employing Theorem 16 the correlation sequence can be represented by the relation

$$\psi_{fg,n} = \sum_{k=0}^{+\infty} f_k g_{k-n} = \lim_{z \to 1+} \sum_{k=0}^{+\infty} f_k g_{k-n} z^{-k} = \lim_{z \to 1+} \mathscr{Z}\{f_k g_{k-n}\}.$$

According to the translation theorem we may write $\mathscr{Z}\{g_{k-n}\} = z^{-n} G(z)$.

(In the definitorical relation of the $\mathscr{Z}$ transform we use the subscript k instead of the usual n.) The transform of the product of the two sequences is represented by the convolution of the transforms

$$\psi_{fg,n} = \lim_{z \to 1+} \frac{1}{2\pi j} \oint_C F(\zeta)\, G\left(\frac{z}{\zeta}\right) \left(\frac{z}{\zeta}\right)^{-n} \frac{1}{\zeta} \, d\zeta =$$

$$= \frac{1}{2\pi j} \oint_C F(\zeta)\, G\left(\frac{1}{\zeta}\right) \zeta^{n-1} \, d\zeta \tag{2.51a}$$

for $n \geqq 0$. Here, integration is carried out along the unit circle as follows from the condition for the radius ϱ in Theorem 21. Further, if the residue theorem is applied we have

$$\psi_{fg,n} = \sum_{i=1}^{K} \operatorname*{res}_{\zeta=\zeta_i} F(\zeta)\, G\left(\frac{1}{\zeta}\right) \zeta^{n-1}, \tag{2.51b}$$

where ζ_i are all the poles of the integrated function inside the circle $|\zeta| = 1$.

Comparing relation (2.51a) with the integral representation of the inverse transform (2.34), we see that the transform of the sequence $\{\psi_{fg,n}\}$ for $n \geqq 0$ is the product of the transforms $F(z)$ and $G(1/z)$, i.e. that we have

$$\mathscr{Z}\{\psi_{fg,n}\} = F(z)\, G\left(\frac{1}{z}\right) = \Psi_{fg}(z) \tag{2.52}$$

for $|z| = 1$.

If we put $f_n = g_n$ for $n \geqq 0$ in our considerations, we arrive at the following representation of the *autocorrelation sequence* of the sequence $\{f_n\}$:

$$\psi_{ff,n} = \psi_n = \sum_{k=0}^{+\infty} f_k f_{k-n} = \frac{1}{2\pi j} \oint_C F(\zeta)\, F\left(\frac{1}{\zeta}\right) \zeta^{n-1} \, d\zeta \tag{2.53a}$$

and, further,

$$\psi_n = \sum_{i=1}^{K} \operatorname*{res}_{\zeta=\zeta_i} F(\zeta)\, F\left(\frac{1}{\zeta}\right) \zeta^{n-1}, \tag{2.53b}$$

Here, ζ_i are all the poles of the integrated function inside the unit circle. Substituting the same equality into relation (2.52) we obtain for the transform of the autocorrelation sequence for $|z| = 1$ the relation

$$\mathscr{Z}\{\psi_n\} = F(z)\, F\left(\frac{1}{z}\right) = \Psi(z) \tag{2.54a}$$

and for the autocorrelation sequence the relation

$$\{\psi_n\} = \mathscr{Z}^{-1}\{\Psi(z)\}\,. \tag{2.54b}$$

This pair of relations is the discrete representation of the *Wiener–Khintchine theorem* with the aid of the one-sided $\mathscr{Z}$ transform.

Further, putting $n = 0$ we obtain the expression for the infinite sum of squares of the terms of the sequence $\{f_n\}$

$$\psi_0 = \sum_{k=0}^{+\infty} f_k^2 = \frac{1}{2\pi j} \oint_C F(\zeta)\, F\left(\frac{1}{\zeta}\right) \zeta^{-1}\, d\zeta\,. \tag{2.55}$$

This equality is the discrete equivalent of the *Parseval theorem* for one-sided sequences.

EXAMPLE. Let us have the sequence $\{e^{\alpha n}\}$ for $\alpha < 0$ and $n \geqq 0$. Its transform for $|z| > |e^{\alpha}|$, where $|e^{\alpha}| < 1$, is

$$\mathscr{Z}\{e^{\alpha n}\} = \frac{z}{z - e^{\alpha}}\,.$$

The transform of the autocorrelation sequence, obtained by (2.54a) in the form

$$\Psi(z) = \frac{z}{z - e^{\alpha}} \frac{\frac{1}{z}}{\frac{1}{z} - e^{\alpha}} = \frac{z}{z - e^{\alpha}} \frac{-e^{-\alpha}}{z - e^{-\alpha}}\,,$$

is a regular function in the annular region $|e^{\alpha}| < |z| < |e^{-\alpha}|$. By the inverse $\mathscr{Z}$ transform, for which we integrate along a circle which lies inside the annulus, we obtain for $n \geqq 0$ the following autocorrelation sequence:

$$\mathscr{Z}^{-1}\{\Psi(z)\} = \frac{1}{2\pi j} \oint_C \frac{-e^{-\alpha} z^n}{(z - e^{\alpha})(z - e^{-\alpha})}\, dz = \frac{e^{\alpha n}}{1 - e^{2\alpha}} = \psi_n\,.$$

If we put $n = 0$, we obtain the sum of squares of the exponential sequence

$$\psi_0 = \sum_{k=0}^{+\infty} e^{2\alpha k} = \frac{1}{1 - e^{2\alpha}}\,.$$

2.4. Two-Sided $\mathscr{Z}$ Transform

The $\mathscr{Z}$ transform introduced in Section 1.2 by the relation

$$F(z) = \sum_{n=0}^{+\infty} f_n z^{-n}$$

was based on the assumption that the sequence $\{f_n\}$ is one-sided, i.e. that we have $f_n = 0$ for $n < 0$. However, in a number of applications in which, e.g., random sequences occur, this condition cannot be fulfilled and it is necessary, therefore, to introduce the more general two-sided $\mathscr{Z}$ transform.

DEFINITION. The *two-sided $\mathscr{Z}$ transform* is defined by the relation

$$F_{\mathrm{II}}(z) = \sum_{n=-\infty}^{+\infty} f_n z^{-n} \tag{2.56}$$

under the assumption that the series is convergent. The sequence is called *$\mathscr{Z}$ transformable* if the series (2.56) converges in a certain annular region.

It is obvious that for the sequence $\{f_n\}$ which satisfies the condition $f_n = 0$ for $n < 0$ we have

$$F_{\mathrm{II}}(z) = \sum_{n=0}^{+\infty} f_n z^{-n} = F(z) .$$

The definitorical relation (2.56) is viewed as decomposed into two infinite sums

$$F_{\mathrm{II}}(z) = \sum_{n=-\infty}^{+\infty} f_n z^{-n} = \sum_{n=0}^{+\infty} f_n z^{-n} + \sum_{n=-1}^{-\infty} f_n z^{-n} ,$$

which define two one-sided transforms

$$F_{\mathrm{II}}(z) = F_{+}(z) + F_{-}(z) , \tag{2.57}$$

where

$$F_{+}(z) = \sum_{n=0}^{+\infty} f_n z^{-n} ,$$

$$F_{-}(z) = \sum_{n=-1}^{-\infty} f_n z^{-n} .$$

The functions $F_{+}(z)$ and $F_{-}(z)$ are identical with the one-sided $\mathscr{Z}$ transforms of the sequence $\{f_n\}$ for $n \geqq 0$ and $n < 0$, respectively. The function $F_{+}(z)$ is called the *right-handed transform*, the function $F_{-}(z)$ is the *left-handed transform.*

For the given relations we shall use the symbolic notation

$$\mathscr{Z}_{\mathrm{II}}\{f_n\} = \mathscr{Z}_+\{f_n\} + \mathscr{Z}_-\{f_n\}.$$

The explicit distinction of the terms "two-sided" and "one-sided" $\mathscr{Z}$ transform will be used only in this subsection, and in the sequel only there where it is necessary to distinguish the two concepts. Otherwise, the term $\mathscr{Z}$ transform will be always understood in the sense of definition (1.1) of Chap. 1.

The definitorical relation of the two-sided $\mathscr{Z}$ transform is a Laurent series of a function defined by its sum. The series is convergent in a certain annulus $1/R_+ < |z| < 1/R_-$ which is simultaneously the domain of regularity of its sum. The division of the two-sided $\mathscr{Z}$ transform to the right-handed and left-handed $\mathscr{Z}$ transforms corresponds (except the assignment of the term f_0) to the division of the Laurent series into the principal and regular parts.

The principal part of the Laurent series

$$F_+(z) = \sum_{n=0}^{+\infty} f_n z^{-n}$$

(in the theory of functions of the complex variable without the term f_0) is convergent in the circle with centre at ∞ and radius of convergence R_+, i.e. it converges in the region $|z| > 1/R_+$. The regular part of the Laurent series

$$F_-(z) = \sum_{n=-1}^{-\infty} f_n z^{-n} = \sum_{n=1}^{+\infty} f_{-n} z^n$$

(in the theory of functions of the complex variable including the term f_0 as well) is a power series which converges in the circle $|z| < R_-$. The

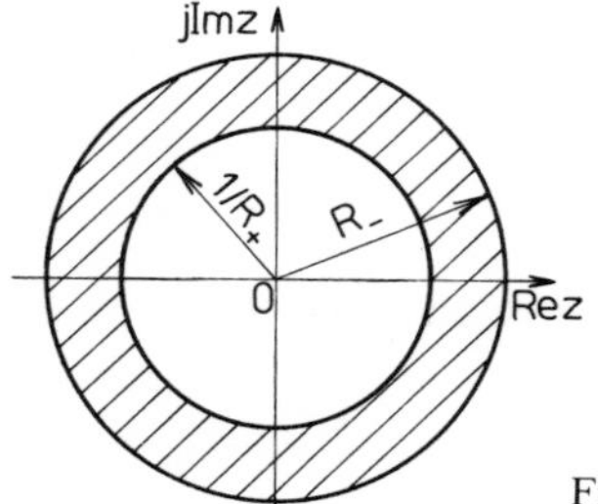

Fig. 12

annular region $1/R_+ < |z| < R_-$ in which the Laurent series $\sum_{n=-\infty}^{+\infty} f_n z^{-n}$ converges is the maximal annular region in which $F_{II}(z)$ is a regular function. It is obvious that the annular region can be extended to a circle with removed centre, $0 < |z| < R_-$, eventually to the exterior of a circle with removed point $z = \infty$, i.e. $1/R_+ < |z| < +\infty$. The annular convergence region of the two-sided $\mathscr{Z}$ transform is indicated in Fig. 12.

The properties of the two-sided $\mathscr{Z}$ transform are given by the properties of Laurent series of functions of complex variables, and they will not be presented here. The reader may also derive them from the properties of the one-sided $\mathscr{Z}$ transform and of the decomposition $F_{II}(z) = F_+(z) + F_-(z)$, where we apply the substitution $z = 1/z'$ for the left-handed transform $F_-(z)$. Now, let us solve examples which will help us comprehend the connection between sequences and their transforms.

EXAMPLE 1. Let us consider the sequence $\{f_n\}$ of Fig. 13 defined by the formula

$$f_n = 1 \qquad \text{for} \qquad n \geqq 0\,,$$
$$f_n = e^{\alpha n} \qquad \text{for} \qquad n < 0, \quad \alpha > 0\,.$$

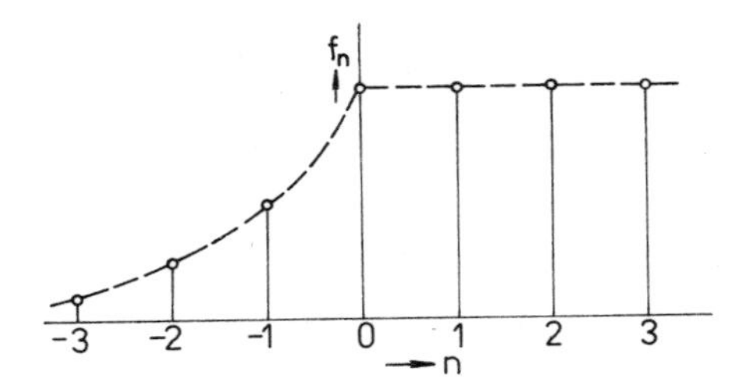

Fig. 13

The application of relation (2.57) yields

$$F_{II}(z) = \sum_{n=0}^{+\infty} z^{-n} + \sum_{n=-1}^{-\infty} e^{\alpha n} z^{-n} = F_+(z) + F_-(z)\,.$$

The first sum defines the right-handed transform of the unit sequence

$$F_+(z) = \frac{1}{1 - z^{-1}} = \frac{z}{z - 1} \qquad \text{for} \qquad |z| > 1 = 1/R_+\,.$$

The second sum is the left-handed transform of the exponential sequence

$$F_{-}(z) = \sum_{n=-1}^{-\infty} e^{\alpha n} z^{-n} = \sum_{n=0}^{+\infty} \left(\frac{z}{e^{\alpha}}\right)^n - 1 = \frac{1}{1 - \dfrac{z}{e^{\alpha}}} - 1 = \frac{-z}{z - e^{\alpha}}$$

which exists for $|z/e^{\alpha}| < 1$, i.e. for $|z| < |e^{\alpha}| = R_{-}$. The resulting transform is

$$F_{\text{II}}(z) = \frac{z}{z-1} + \frac{-z}{z-e^{\alpha}} = \frac{(1-e^{\alpha})\,z}{(z-1)(z-e^{\alpha})}\,.$$

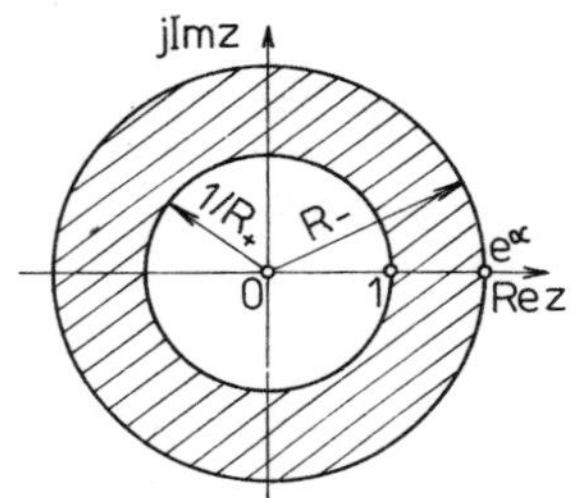

Fig. 14

Fig. 14 shows the convergence regions of the partial sums and the zero and poles of the resulting transform. We see that the hatched annular region is the common convergence region of the individual transforms. Consequently, it is also the region of regularity of the function which defines the two-sided $\mathscr{Z}$ transform.

EXAMPLE 2. To elucidate the relationship between transforms and sequences in the case of the two-sided $\mathscr{Z}$ transform we determine the transform of a sequence symmetrical with respect to the vertical axis (see Fig. 15), i.e. the sequence $\{f_n\}$ defined by the relations

$$f_n = e^{-\alpha n} \quad \text{for} \quad n \geqq 0, \quad \alpha > 0\,,$$
$$f_n = 1 \quad \text{for} \quad n < 0\,.$$

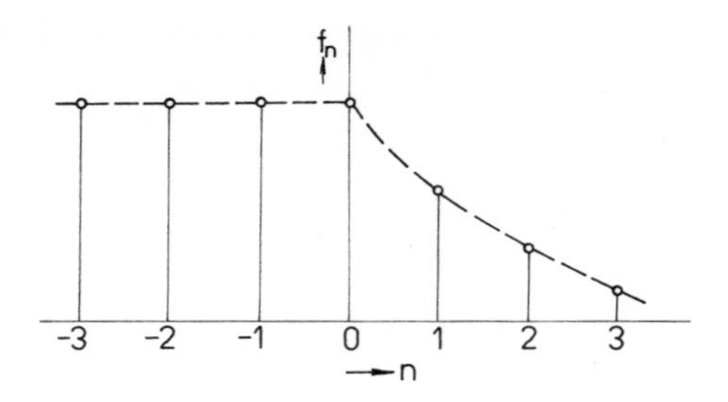

Fig. 15

Then we have

$$F_{II}(z) = F_+(z) + F_-(z) = \sum_{n=0}^{+\infty} e^{-\alpha n} z^{-n} + \sum_{n=-1}^{-\infty} z^{-n} =$$

$$= \frac{z}{z - e^{-\alpha}} + \frac{-z}{z - 1} = \frac{(e^{-\alpha} - 1)\, z}{(z - e^{-\alpha})(z - 1)}.$$

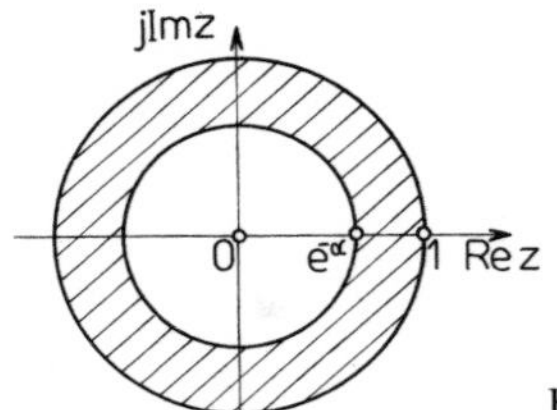

Fig. 16

The series which defines $F_{II}(z)$ converges in the annular region $|e^{-\alpha}| < |z| < 1$ which is identical with the region of regularity of the transform. The region of convergence, the zero and the poles are shown in Fig. 16.

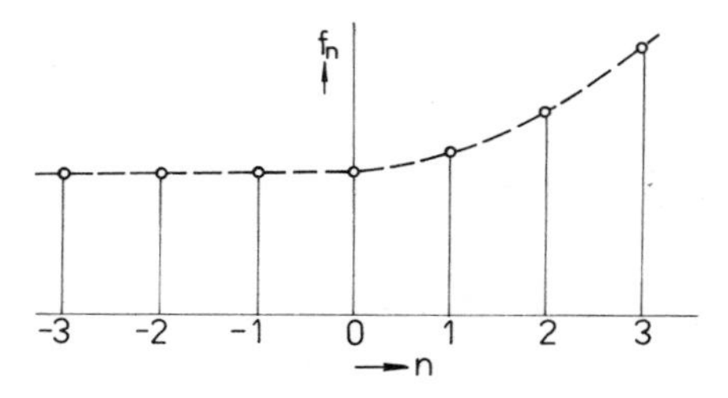

Fig. 17

EXAMPLE 3. To make clear the difference between the one-sided and two-sided $\mathscr{Z}$ transforms we employ the sequence $\{f_n\}$ given by the formula (see Fig. 17)

$$f_n = e^{\alpha n} \quad \text{for} \quad n \geqq 0, \quad \alpha > 0,$$
$$f_n = 1 \quad \text{for} \quad n < 0.$$

The right-handed transform exists for $|z| > |e^\alpha| > 1$

$$F_+(z) = \frac{z}{z - e^\alpha},$$

while the left-handed transform exists for $|z| < 1$

$$F_-(z) = \frac{-z}{z - 1}.$$

In Fig. 18, the convergence regions, poles and zeros of the partial transforms are shown. From the figure it is obvious that the convergence regions of the transforms $F_+(z)$ and $F_-(z)$ do not intersect. Therefore, the existence of the partial transforms does not guarantee the existence of the two-sided $\mathscr{Z}$ transform $F_{II}(z)$.

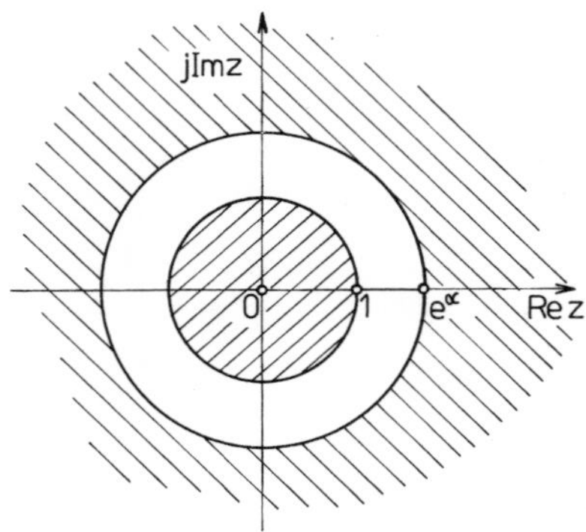

Fig. 18

However, let us ask the question to what does the function obtained as the sum

$$\hat{F}(z) = \frac{z}{z - e^\alpha} - \frac{z}{z - 1} = \frac{z(e^\alpha - 1)}{(z - e^\alpha)(z - 1)}$$

correspond. The function $\hat{F}(z)$ is regular in three domains:

1. $|z| > |e^{\alpha}| > 1$;
2. $|z| < 1$;
3. $1 < |z| < |e^{\alpha}|$.

In the first case, it is possible to consider the sum to be the right-handed transform $\hat{F}(z) = \hat{F}_{+}(z)$, and with the aid of Theorem 19 to determine, for $\varrho > 1/R = |e^{\alpha}|$, the sequence

$$\hat{f}_n = e^{\alpha n} - 1^n \qquad \text{for} \qquad n \geqq 0 .$$

The sequence is indicated in the right part of Fig. 19.

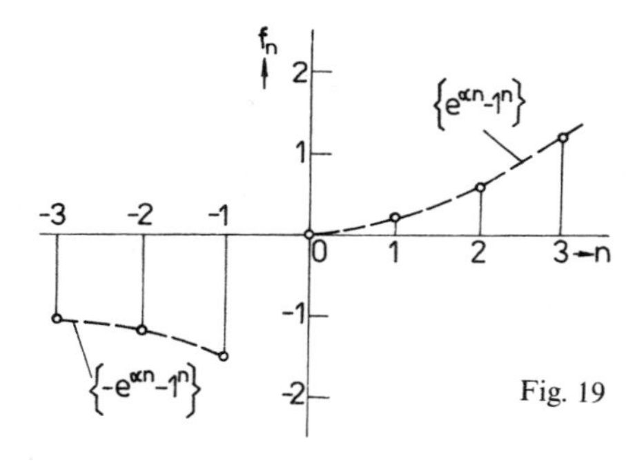

Fig. 19

In the second case, the function $\hat{F}(z)$ is the left-handed transform $\hat{F}(z) = \hat{F}_{-}(z)$, and it is possible to expand it into a power series of the variable z. One obtains

$$\hat{F}_{-}(z) = - \sum_{n=-1}^{-\infty} e^{\alpha n} z^{-n} - \sum_{n=-1}^{-\infty} z^{-n} = - \sum_{n=-1}^{-\infty} (e^{\alpha n} + 1^n)\, z^{-n} .$$

For the sequence we thus have

$$\hat{f}_n = -(e^{\alpha n} + 1^n) \qquad \text{for} \qquad n < 0 .$$

This sequence is indicated in the left part of Fig. 19.

In the third case, the function $\hat{F}(z)$ is regular in the annular region and it can thus be viewed as the two-sided $\mathscr{Z}$ transform: $\hat{F}(z) = \hat{F}_{II}(z)$. Comparison with the result of Example 1 shows that the two functions differ by their signs only. Thus, the resulting sequence is given by the formula

$$f_n = -1^n \quad \text{for} \quad n \geqq 0,$$
$$f_n = -e^{\alpha n} \quad \text{for} \quad n < 0.$$

From the example the relationship of the region of regularity of the transform with the form of the object function is obvious, i.e. the connection with the sequence $\{f_n\}$. In various applications it is necessary to choose a suitable transformation, with account taken of the physical interpretation.

2.4.1. *Inverse Two-Sided $\mathscr{Z}$ Transform*

By the *inverse two-sided $\mathscr{Z}$ transform* we understand – similarly as in the case of the one-sided $\mathscr{Z}$ transform – the determination of the sequence $\{f_n\}$ to a given transform $F_{II}(z)$. We use the symbolic notation

$$\{f_n\} = \mathscr{Z}_{II}^{-1}\{F_{II}(z)\}.$$

Again, the inverse transform can be obtained in several ways. The simplest procedure consists in the decomposition of the given two-sided transform to partial simple one-sided transforms for which the object functions are known, or for which the sequences are determined by one-sided inverse transforms. We exploit with advantage the uniqueness of the $\mathscr{Z}$ transform which follows from the uniqueness of the Laurent expansion of a function of the complex variable.

The general approach to the inverse two-sided $\mathscr{Z}$ transform consists again in the application of the Cauchy integral theorem to the determination of the coefficients of the Laurent expansion.

THEOREM 22. *Let $F_{II}(z)$ be a regular function in the annular region $1/R_+ < |z| < R_-$. Then a unique sequence $\{f_n\}$ exists for which we have $\mathscr{Z}_{II}\{f_n\} = F_{II}(z)$, namely*

$$f_n = \frac{1}{2\pi j} \oint_C F_{II}(z)\, z^{n-1}\, dz \quad \textit{for} \quad n = \ldots, -2, -1, 0, 1, 2, \ldots. \tag{2.58}$$

We integrate along the circle C for which we have $z = \varrho\, e^{j\varphi}$, $1/R_+ < \varrho < R_-$, and $0 \leqq \varphi \leqq 2\pi$.

For the evaluation of the integral we use the residue theorem, as a rule. Then

$$f_n = \sum_{i=1}^{K} \operatorname*{res}_{z=z_i} F_{II}(z)\, z^{n-1}\,, \tag{2.59}$$

where K is the number of different poles z_i of the function $F_{II}(z)\, z^{n-1}$ which lie in the circle C with radius $\varrho > 1/R_+$ (see Fig. 20).

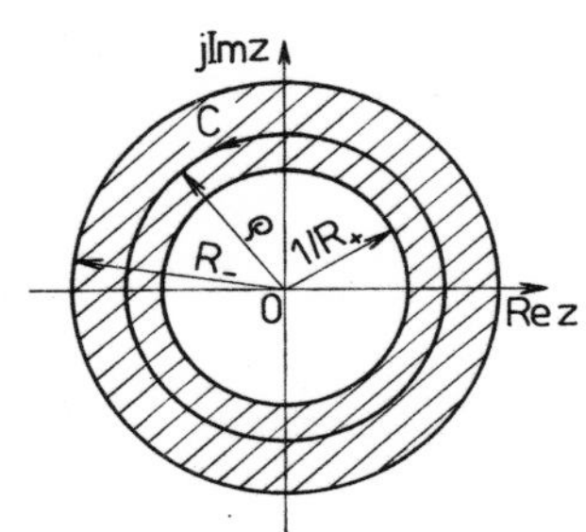

Fig. 20

For our further considerations let us assume that the transform $F_{II}(z)$ has no zero, nor pole, nor any other singularity at the point $z = 0$. (This condition represents no restriction, since every function with either a zero or a pole at the origin may be considered to be the transform of a translated sequence.) For $n > 0$, the poles z_i are given only by poles of the transform $F_{II}(z)$ which lie inside the circle with radius ϱ. For $n \leqq 0$, these poles will be joined by a pole at the origin of order $m = 1 - n$. If we use for the evaluation of the residue of the function $F_{II}(z)\, z^{n-1}$ at the point $z = 0$ the relation for the residue at the pole (2.36) with multiplicity m, we obtain for $n \leqq 0$ the equation

$$\operatorname*{res}_{z=0} F_{II}(z)\, z^{n-1} = \frac{1}{(-n)!} \frac{d^{-n}}{dz^{-n}} F_{II}(z)\bigg|_{z=0}.$$

Then the resulting relation for the terms of the sequence $\{f_n\}$ for $n \leqq 0$ is

$$f_n = \sum_{i=1}^{K} \operatorname*{res}_{z=z_i} F_{II}(z)\, z^{n-1} + \frac{1}{(-n)!} \frac{d^{-n}}{dz^{-n}} F_{II}(z)\bigg|_{z=0}. \tag{2.60}$$

EXAMPLE 1. We are given the two-sided transform of the sequence $\{f_n\}$

$$F_{\mathrm{II}}(z) = \frac{z}{(z-1)(z-2)}.$$

$F_{\mathrm{II}}(z)$ is a regular function in the region $1 < |z| < 2$. By Theorem 22 we have

$$f_n = \frac{1}{2\pi \mathrm{j}} \oint_C \frac{z^n}{(z-1)(z-2)} \,\mathrm{d}z\,,$$

where we integrate along the circle with radius $1 < \varrho < 2$. For $n \geqq 0$, the integrated function has a single pole in the interior of this circle at the point $z = 1$, and we have

$$f_n = -1^n \qquad \text{for} \qquad n \geqq 0\,.$$

For $n < 0$, moreover, the integrated function possesses a pole of nth order at the origin and we have

$$f_n = \operatorname*{res}_{z=1} \frac{1}{(z-1)(z-2)\,z^{-n}} + \operatorname*{res}_{z=0} \frac{1}{(z-1)(z-2)\,z^{-n}} =$$

$$= -1^n + \frac{1}{(-n-1)!} \lim_{z \to 0} \frac{\mathrm{d}^{-n-1}}{\mathrm{d}z^{-n-1}} \frac{1}{(z-1)(z-2)}.$$

Hence we obtain, by successive evaluation,

$$f_{-1} = -1 + \frac{1}{2} = -\frac{1}{2},$$

$$f_{-2} = -1 + \frac{3}{4} = -\frac{1}{4}, \;\ldots\,.$$

When executing the inverse transformation for $n \leqq 0$, it is possible to avoid the evaluation of the derivatives of nth order if we introduce the substitution $z = 1/\zeta$, for the determination of the terms of f_n for $n \leqq 0$, when computing the integral. We obtain

$$f_n = \frac{1}{2\pi \mathrm{j}} \oint_C F_{\mathrm{II}}(z)\, z^{n-1} \,\mathrm{d}z =$$

$$= -\frac{1}{2\pi \mathrm{j}} \oint_{C'} F_{\mathrm{II}}\!\left(\frac{1}{\zeta}\right) \frac{1}{\zeta^{n-1}} \frac{1}{\zeta^2} \,\mathrm{d}\zeta =$$

$$= -\frac{1}{2\pi \mathrm{j}} \oint_{C'} F_{\mathrm{II}}\!\left(\frac{1}{\zeta}\right) \frac{1}{\zeta^{n+1}} \,\mathrm{d}\zeta\,,$$

where we integrate along the circle C' with radius $\varrho' = 1/\varrho$ in the plane ζ in the negative sense. By a change of the direction of integration we obtain the final relation for the determination of the terms f_n for $n \leqq 0$ as follows:

$$f_n = \frac{1}{2\pi j} \oint_{C'} F_{II}\left(\frac{1}{\zeta}\right) \frac{1}{\zeta^{n+1}} \, d\zeta . \tag{2.61}$$

Thus, we integrate along the circle C' inside which lie the poles of the function $F_{II}(1/\zeta)$ and outside which lie the poles of the function $F_{II}(z)$. If we apply to this integral the residue theorem, then we will have for $n \leqq 0$

$$f_n = \sum_{i=1}^{K_1} \operatorname*{res}_{\zeta = \zeta_i} F_{II}\left(\frac{1}{\zeta}\right) \frac{1}{\zeta^{n+1}}, \tag{2.62}$$

where K_1 is the number of distinct poles ζ_i of the function $F_{II}(1/\zeta)\, \zeta^{-(n+1)}$ which lie in the circle C' with radius $1/R_- < \varrho' < R_+$.

EXAMPLE 2. Applying the approach described above we solve once more the second part of Example 1, i.e. we determine the terms f_n for $n < 0$.

Upon substitution into (2.61) we have, for $n < 0$,

$$f_n = \frac{1}{2\pi j} \oint_{C'} \frac{1/\zeta}{(1/\zeta - 1)(1/\zeta - 2)} \frac{1}{\zeta^{n+1}} \, d\zeta =$$
$$= \frac{1}{2\pi j} \oint_{C'} \frac{1/2}{(\zeta - 1)(\zeta - 1/2)} \frac{1}{\zeta^n} \, d\zeta ,$$

where we integrate along the circle C' with radius $1/2 < \varrho' < 1$. With the aid of the residue theorem we obtain, for $n < 0$,

$$f_n = \operatorname*{res}_{\zeta = 1/2} \frac{1}{2} \frac{1}{(\zeta - 1)(\zeta - 1/2)\, \zeta^n} = -(2^n) .$$

The resulting sequence $\{f_n\}$ is thus determined by the relations

$$f_n = -(1^n) \quad \text{for} \quad n \geqq 0 ,$$
$$f_n = -(2^n) \quad \text{for} \quad n < 0 .$$

2.4.2. *Inverse Two-Sided Transform with the Aid of Differentiation*

In Subsection 2.4.1, the object function $\{f_n\}$ was determined to the two-sided transform $F_{II}(z)$ with the aid of integration. However, the object function $\{f_n\}$ may be also determined employing the one-sided transforms obtained in Section 2.4 by the decomposition of the transform $F_{II}(z)$ which is regular in the annular region $1/R_+ < |z| < R_-$:

$$F_{II}(z) = F_+(z) + F_-(z),$$

where

$$F_+(z) = \sum_{n=0}^{+\infty} f_n z^{-n} \qquad \text{for} \qquad |z| > 1/R_+$$

and

$$F_-(z) = \sum_{n=-1}^{-\infty} f_n z^{-n} \qquad \text{for} \qquad |z| < R_- .$$

For the inverse transform of $F_+(z)$ with the aid of differentiation Theorem 20 holds which enables the determination of the terms of the sequence $\{f_n\}$ for $n \geqq 0$. This theorem is also valid for $F_-(z)$ if the substitution $\zeta = 1/z$ is used. Thus, it is possible to evaluate also the terms of the sequence $\{f_n\}$ for $n < 0$. This procedure of the inverse transform of $F_{II}(z)$ which employs differentiation is summarized in the following theorem:

THEOREM 23. *Let $F_{II}(z)$ be a function which is regular in the annular region $1/R_+ < |z| < R_-$. If the decomposition $F_{II}(z) = F_+(z) + F_-(z)$ is performed so that $F_+(z)$ is regular in the region $|z| > 1/R_+$ while $F_-(z)$ is regular in the circle $|z| < R_-$, then the terms of the sequence $\{f_n\}$ for which we have $\mathscr{Z}_{II}\{f_n\} = F_{II}(z)$ are given by the relations*

$$f_n = \frac{1}{n!} \frac{\mathrm{d}^n}{\mathrm{d}\zeta^n} F_+\left(\frac{1}{\zeta}\right)\bigg|_{\zeta=0} \qquad \textit{for} \qquad n \geqq 0, \tag{2.63a}$$

$$f_n = \frac{1}{(-n)!} \frac{\mathrm{d}^{-n}}{\mathrm{d}z^{-n}} F_-(z)\bigg|_{z=0} \qquad \textit{for} \qquad n < 0. \tag{2.63b}$$

To the transform $F_+(z)$ in the form of a rational function it is possible to determine the sequence $\{f_n\}$ for $n \geqq 0$ by recurrent computations – without evaluating the derivatives – in line with the procedure discussed in Subsection 2.2.2. This consists in the division of a polynomial by a polynomial. Similarly it is also possible to proceed when executing the inverse transform of $F_-(z)$. However, $F_-(z)$ has to be expanded into a series of the variable z:

$$F_-(z) = \sum_{n=-1}^{-\infty} f_n z^{-n} = \sum_{n=1}^{+\infty} f_{-n} z^n .$$

The application of the procedure discussed in Subsection 2.2.2 asks for the introduction of the substitution $z = 1/\zeta$ in the function $F_-(z)$. The required form of the expansion assumes that the function $F_-(z)$ has a zero at the origin. The resulting procedure is summarized as follows:

Assume that the function $F_-(z)$ is of the form

$$F_-(z) = \frac{a_0 z^s + a_1 z^{s-1} + \ldots + a_{s-1} z}{b_0 z^s + b_1 z^{s-1} + \ldots + b_{s-1} z + 1},$$

where the absolute term in the numerator vanishes while the absolute term in the denominator is modified by suitable normalization to the unit. Let us divide the polynomial in the numerator by the polynomial in the denominator starting from the absolute terms:

$$\begin{array}{l}
(a_{s-1} z + \quad a_{s-2} z^2 + \ldots + a_0 z^s) : (1 + b_{s-1} z + \ldots + b_1 z^{s-1} + b_0 z^s) = \\
\underline{\pm a_{s-1} z \pm b_{s-1} a_{s-1} z^2 + \ldots \pm \qquad b_1 a_{s-1} z^s \pm b_0 a_{s-1} z^{s+1}} \\
(a_{s-2} - b_{s-1} a_{s-1}) z^2 + \ldots + (a_0 - b_1 a_{s-1}) z^s - b_0 a_{s-1} z^{s+1} \\
\quad \vdots \qquad\qquad\qquad\qquad \vdots \qquad\qquad \vdots \\
\qquad\qquad\qquad = a_{s-1} z + (a_{s-2} - b_{s-1} a_{s-1}) z^2 + \ldots .
\end{array}$$

The comparison of this ratio with the expansion of $F_-(z)$ implies that $f_{-1} = a_{s-1}$, $f_{-2} = a_{s-2} - b_{s-1} a_{s-1}, \ldots$. The negative subscripts of the terms of the sequence are the exponents of the variable z in the expansion of the function $F_-(z)$. The computation corresponds to the evaluation of the recurrent relation

$$f_{-n} = a_{s-n} - \sum_{i=1}^{s} b_{s-i} f_{-n+i}, \tag{2.64}$$

where $a_{s-n} = a_{s-i}$ for $i = n = 1, 2, \ldots, s$, $a_{s-n} = 0$ for $n > s$. Substitution into the recurrent relation is facilitated again by the strip method:

Numerator	$\{f_n\}$
0	$f_0 = 0$
a_{s-1}	$f_{-1} = a_{s-1}$
a_{s-2}	$f_{-2} = a_{s-2} - b_{s-1}f_{-1}$
$\vdots$	$\vdots$
a_0	$f_{-s} = a_0 - \sum_{i=1}^{s} b_{s-i}f_{-s+i}$
	$\vdots$
	$f_{-n} = -\sum_{i=1}^{s} b_{s-i}f_{-n+i}$

Denominator
$-b_0$
$-b_1$
$\vdots$
$-b_{s-2}$
$-b_{s-1}$
$\longleftarrow$

Strip of paper

EXAMPLE. The application of the algebraic approach to the inverse two-sided transform is demonstrated by the case of the transform

$$F_{\text{II}}(z) = \frac{z}{(z-1)(z-2)},$$

which is a regular function in the annular region $1 < |z| < 2$. The transform is decomposed into partial fractions

$$F_{\text{II}}(z) = \frac{-z}{z-1} + \frac{z}{z-2} = F_+(z) + F_-(z).$$

The terms of the sequence $\{f_n\}$ for $n \geqq 0$ are obtained by the expansion of the function

$$F_+(z) = \frac{-z}{z-1} = -\sum_{n=0}^{+\infty} z^{-n} \qquad \text{for} \qquad |z| > 1.$$

For the computation of the terms for $n < 0$ we apply to the expansion of the function

$$F_-(z) = \frac{z}{z-2} = \frac{-0.5z}{-0.5z+1} \qquad \text{for} \qquad |z| < 2$$

the simplified strip method as follows:

Numerator	$\{f_n\}$
0	$f_0 \;\;= 0$
-0.5	$f_{-1} = -0.5$
	$f_{-2} = -0.25$
	$f_{-3} = -0.125$
	$\vdots$

Denominator
0.5
⟵

Consequently, the object function to the two-sided transform $F_{\text{II}}(z)$ is given by the formula

$$f_n = -(1)^n \qquad \text{for} \qquad n \geqq 0\,,$$
$$f_n = -(0.5^{-n}) = -(2)^n \qquad \text{for} \qquad n < 0\,.$$

Moreover, let us present the algebraic procedure for the two-sided inverse $\mathscr{Z}$ transform of $G_{\text{II}}(z)$ when the transform is given in the form of a product

$$G_{\text{II}}(z) = F_+(z)\,G_-(z)\,.$$

Let the transform $F_+(z)$ be given by a polynomial of Nth degree of the variable z^{-1}

$$F_+(z) = \sum_{n=0}^{N} f_n z^{-n}\,,$$

and let $G_-(z)$ be given by a rational function of the variable z

$$G_-(z) = \frac{a_0 z^s + a_1 z^{s-1} + \ldots + a_{s-1} z}{b_0 z^s + b_1 z^{s-1} + \ldots + b_{s-1} z + 1}$$

which is regular in the circle $|z| < R_-$. We proceed similarly as in the case of expansion (b) in Subsection 2.2.2. – we use the substitution $z = 1/\zeta$. The substitution causes that the sequence $\{f_n\}$ is processed not starting with the term f_0 but from the end, i.e. starting from the term f_N, according to the recurrent relation

$$g_n = \sum_{i=0}^{s-1} a_i f_{n+s-i} - \sum_{i=0}^{s-1} b_i g_{n+s-i} \tag{2.65}$$

where $n = N, N-1, N-2, \ldots$, $g_n = f_n = 0$ for $n > N$, and $f_n = 0$ for $n < 0$. The transform of the sequence $\{g_n\}$ represented by a series is

$$G_{\mathrm{II}}(z) = \mathscr{Z}_{\mathrm{II}}\{g_n\} = \sum_{n=-\infty}^{N-1} g_n z^{-n}.$$

The sequence $\{g_n\}$ is evaluated with the aid of the strip method again:

Paper template	$\{f_n\}$	$\{g_n\}$	
Numerator	f_N	$g_N = 0$	Denominator
	f_{N-1}	$g_{N-1} = a_{s-1} f_N$	
	f_{N-2}	$g_{N-2} = a_{s-1} f_{N-1} + a_{s-2} f_N - b_{s-1} g_{N-1}$	
a_0	$\vdots$	$\vdots$	$-b_0$
$\vdots$	f_0	g_0	$\vdots$
a_{s-2}		$\vdots$	$-b_{s-2}$
a_{s-1}			$-b_{s-1}$
$0 \rightarrow$		$g_k = \sum_{i=0}^{s-1} a_i f_{k+s-i} - \sum_{i=0}^{s-1} b_i g_{k+s-i}$	$\leftarrow$

Chapter 3

Application of the $\mathscr{Z}$ Transform to the Analysis of Linear Discrete Systems

3.1. Solution of Difference Equations

Difference equations express dependences between sequences and their differences. As far as such a dependence is linear, we speak of a *linear difference equation.* A linear difference equation of order s is written in the form

$$B_s\Delta^s h_n + B_{s-1}\Delta^{s-1}h_n + \dots + B_0 h_n = \\ = A_s\Delta^s f_n + A_{s-1}\Delta^{s-1}f_n + \dots + A_0 f_n , \tag{3.1}$$

where $B_0, B_1, \dots, B_s$ and $A_0, A_1, \dots, A_s$ are coefficients of the difference equation, $\{f_n\}$ is a known sequence, and $\{h_n\}$ is the desired solution of the difference equation. A difference equation with non-zero right-hand side is called *nonhomogeneous,* a difference equation with $f_n \equiv 0$ is called *homogeneous.*

For the differences of the sequence $\{f_n\}$ we have

$$\begin{aligned} \Delta^0 f_n &= f_n , \\ \Delta^1 f_n &= f_{n+1} - f_n , \\ \Delta^2 f_n &= \Delta f_{n+1} - \Delta f_n = f_{n+2} - 2f_{n+1} + f_n , \\ &\vdots \\ \Delta^s f_n &= \Delta^{s-1} f_{n+1} - \Delta^{s-1} f_n = \sum_{i=0}^{s} (-1)^i \binom{s}{i} f_{n+s-i} , \end{aligned} \tag{3.2}$$

where

$$\binom{s}{i} = \frac{s!}{i!(s-i)!}$$

are binomial coefficients. Similar relations hold for the sequence $\{h_n\}$. If the differences in difference equation (3.1) are expressed using relation (3.2), we obtain the following equivalent form of the difference equation which is more useful for a number of applications:

$$b_s h_{n+s} + b_{s-1} h_{n+s-1} + \dots + b_0 h_n \\ = a_s f_{n+s} + a_{s-1} f_{n+s-1} + \dots + a_0 f_n . \tag{3.3}$$

If the initial values $h_0, h_1, \ldots, h_{s-1}$ are given, it is possible to determine with the aid of equation (3.3) the other terms of the sequence, i.e. h_n for $n = s, s+1, \ldots$, by recurrent computation. The equation in the form (3.3) is therefore called a *recurrent equation* of order s.

Difference equations are solved either by methods of difference calculus or using functional transforms – either the Laplace transform or the $\mathscr{Z}$ transform. The solution of a difference equation with the aid of the $\mathscr{Z}$ transform is illustrated below using an example of a recurrent equation of order 2.

Let a discrete system be given whose behavior is described by a difference equation of order 2 with constant coefficients and right-hand side

$$h_{n+2} + b_1 h_{n+1} + b_0 h_n = a_2 f_{n+2} + a_1 f_{n+1} + a_0 f_n . \tag{3.4}$$

The initial conditions are given by the terms h_0, h_1 and f_0, f_1.

Assume that the two sequences are of the exponential type, thus being $\mathscr{Z}$ transformable, and that we have $\mathscr{Z}\{f_n\} = F(z)$ and $\mathscr{Z}\{h_n\} = H(z)$. Applying the translation theorem we obtain the transform of the recurrent equation

$$\begin{aligned} &z^2 H(z) - z^2 h_0 - z h_1 + b_1[z H(z) - z h_0] + b_0 H(z) = \\ &= a_2[z^2 F(z) - z^2 f_0 - z f_1] + a_1[z F(z) - z f_0] + a_0 F(z) . \end{aligned} \tag{3.5}$$

The transform of the complete solution of the nonhomogeneous difference equation reduces to

$$\begin{aligned} H(z) = &\frac{a_2 z^2 + a_1 z + a_0}{z^2 + b_1 z + b_0} F(z) + \\ &+ \frac{z^2(h_0 - a_2 f_0) + z(h_1 + b_1 h_0 - a_1 f_0 - a_2 f_1)}{z^2 + b_1 z + b_0} . \end{aligned} \tag{3.6}$$

The object function, i.e. the sequence $\{h_n\}$, is obtained by the inverse $\mathscr{Z}$ transform. Three approaches given in Section 2.2 can be used: decomposition into partial fractions and the application of the list of transforms, the approach based on the application of the inverse transform integral, or recurrent computations. As a rule, however, only the first two approaches are being used for the solution of difference equations since the third approach is practically equivalent with the recurrent process of the solution of a difference equation – this does not lead to a functional formula for the sequence $\{h_n\}$ in most cases.

Now, let us assume that the system described by a difference equation of order 2 is at rest for $n < 0$, i.e. assume that $f_n = h_n = 0$ for $n < 0$. With the aid of the known initial elements of the sequence $\{f_n\}$ we obtain

$$h_0 = a_2 f_0 , \qquad h_1 = a_1 f_0 + a_2 f_1 - b_1 h_0$$

from the difference equation (3.4) for $n = -2, -1$. Upon substitution into the transform of the complete solution we see that the transform simplifies into the form

$$H_0(z) = \frac{a_2 z^2 + a_1 z + a_0}{z^2 + b_1 z + b_0} F(z) . \tag{3.7}$$

Hence, the *particular solution* $\{h_{0,n}\}$ of the given difference equation is obtained by the inverse transform.

On the other hand, if $\{f_n\} \equiv 0$ and the system described by the difference equation is not at rest for $n < 0$, we have here the solution of a homogeneous difference equation under nonzero initial conditions. We then have

$$H_1(z) = \frac{z^2 h_0 + z(h_1 + b_1 h_0)}{z^2 + b_1 z + b_0} \tag{3.8}$$

and $\{h_{1,n}\} = \mathscr{Z}^{-1}\{H_1(z)\}$. The transform of the complete solution of the difference equation is obviously given as the sum of the transforms of the two solutions

$$H(z) = H_0(z) + H_1(z) . \tag{3.9}$$

EXAMPLE. The difference equation of order 2

$$h_{n+2} + b_1 h_{n+1} + b_0 h_n = f_n$$

is given, where the sequence $\{f_n\}$ on the right-hand side is a unit sequence

$$f_n = 1 \quad \text{for} \quad n \geqq 0 ,$$
$$f_n = 0 \quad \text{for} \quad n < 0 .$$

Then the transform of the complete solution assumes the form

$$H(z) = \frac{1}{z^2 + b_1 z + b_0} \, \frac{z}{z - 1} + \frac{z(z h_0 + h_1 + b_1 h_0)}{z^2 + b_1 z + b_0} .$$

To enable the application of the theorem on the inverse transform we find the poles of the function $H(z)$. One pole lies at the point $z = 1$, the other two are given by the solution of the quadratic equation

$$z^2 + b_1 z + b_0 = 0\,.$$

Denote the roots of this equation by $z_{1,2}$. Assume that $z_1 \neq z_2$, i.e. that the poles of the function $G(z)$ are simple. Then the transform can be written in the form

$$H(z) = \frac{z}{(z-1)(z-z_1)(z-z_2)} + \frac{z(zh_0 + b_1 h_0 + h_1)}{(z-z_1)(z-z_2)}\,.$$

The transform of the particular solution $H_0(z)$ of the difference equation is obtained by substituting the zero initial conditions, in our case $h_0 = h_1 = 0$:

$$H_0(z) = \frac{z}{(z-1)(z-z_1)(z-z_2)}\,.$$

To establish the inverse transform we apply the residue theorem (2.35) which yields, for simple poles, the relation

$$h_{0,n} = \mathscr{Z}^{-1}\{H_0(z)\} = \frac{1}{(1-z_1)(1-z_2)} + \\ + \frac{z_1^n}{(z_1-1)(z_1-z_2)} + \frac{z_2^n}{(z_2-1)(z_2-z_1)} = \\ = \frac{1}{(1-z_1)(1-z_2)(z_1-z_2)} \cdot \\ \cdot \left[z_1 - z_2 - (1-z_2)\,z_1^n + (1-z_1)\,z_2^n\right]$$

for $n \geqq 0$. It is not immediately obvious that the obtained solution satisfies the initial conditions. However, substituting $n = 0$ and $n = 1$ we verify that the sequence $\{h_{0,n}\}$ satisfies the conditions.

The transform of the solution of the homogeneous equation is obtained for $f_n \equiv 0$ and nonzero initial conditions h_0 and h_1:

$$H_1(z) = \frac{z(zh_0 + h_1 + b_1 h_0)}{(z-z_1)(z-z_2)}\,.$$

In the list of transforms we find that

$$\mathscr{Z}^{-1}\left\{\frac{z}{(z - z_1)(z - z_2)}\right\} = \frac{1}{z_1 - z_2}(z_1^n - z_2^n).$$

We substitute and then apply the translation theorem. For $n \geqq 0$ we thus obtain

$$\begin{aligned} h_{1,n} = \mathscr{Z}^{-1}\{H_1(z)\} &= \frac{h_0}{z_1 - z_2}(z_1^{n+1} - z_2^{n+1}) + \\ &+ \frac{b_1 h_0 + h_1}{z_1 - z_2}(z_1^n - z_2^n) = \\ &= \frac{1}{z_1 - z_2}\{z_1^n[h_0(z_1 + b_1) + h_1] - \\ &- z_2^n[h_0(z_2 + b_1) + h_1]\}. \end{aligned}$$

The obtained sequence $\{h_{1,n}\}$ satisfies the initial conditions. This can be verified by substituting $n = 0, 1$ and $b_1 = -(z_1 + z_2)$.

The complete solution of the difference equation is given as the sum of the particular solution and the solution of the homogeneous equation

$$\{h_n\} = \{h_{0,n}\} + \{h_{1,n}\}.$$

The described approach to the solution of a second-order difference equation can also be applied to the solution of higher order difference equations. However, in most cases it will be necessary to solve an algebraic equation of higher order and the resulting formula will be more complicated.

In some applications we encounter difference equations of the type

$$\begin{aligned} &b_s h(t + sT) + b_{s-1} h[t + (s - 1)T] + \ldots + b_0 h(t) = \\ &= a_s f(t + sT) + \ldots + a_0 f(t), \end{aligned} \tag{3.10}$$

where t is a continuous real variable. To be able to solve even such equations with the aid of the $\mathscr{Z}$ transform, for all t from the interval $0 \leqq t < +\infty$, we put $t = (n + \varepsilon)T$, $n \geqq 0$ integer, $T > 0$, $0 \leqq \varepsilon < 1$. This reduces the functions $f(t)$ and $h(t)$ to sequences of functions of the variable ε for which we have

$$\begin{aligned} f_n(\varepsilon) &= f(t)|_{t=(n+\varepsilon)T}, \\ h_n(\varepsilon) &= h(t)|_{t=(n+\varepsilon)T}. \end{aligned} \tag{3.11}$$

Further, the same procedure is applied as used for the difference equation of the form (3.3). Again, we indicate the approach using the difference equation of order 2

$$h(t + 2T) + b_1 h(t + T) + b_0 h(t) = f(t) . \tag{3.12}$$

Substitution of relations (3.11) into the equation yields the following recurrent equation for the sequence of functions $\{h_n(\varepsilon)\}$

$$h_{n+2}(\varepsilon) + b_1 h_{n+1}(\varepsilon) + b_0 h_n(\varepsilon) = f_n(\varepsilon) , \tag{3.13}$$

to which the $\mathscr{Z}$ transform is applied. We introduce the notation $\mathscr{Z}\{f_n(\varepsilon)\} = F(z, \varepsilon)$, $\mathscr{Z}\{g_n(\varepsilon)\} = G(z, \varepsilon)$. The transform of the complete solution of the equation then becomes

$$H(z, \varepsilon) = \frac{F(z, \varepsilon)}{z^2 + b_1 z + b_0} + \frac{z[z h_0(\varepsilon) + b_1 h_0(\varepsilon) + h_1(\varepsilon)]}{z^2 + b_1 z + b_0} . \tag{3.14}$$

The inverse transform yields the sequence of functions $\{h_n(\varepsilon)\}$, for which the subscript n, $n = 0, 1, 2, \ldots$, and the variable $0 \leqq \varepsilon < 1$ determine entirely the behaviour of the function $h(t)$ for $0 \leqq t < +\infty$. The application of this approach will be thoroughly discussed later in Section 3.4.

3.1.1. *Summation of Series*

To determine the formula for the sum of a series it is possible to use different approaches which belong, e.g., to the general theory of series, the Fourier series theory, or which consist in the solution of differential equations or in the application of integral functional transforms.

Summation of series is also a part of difference calculus. The problem of determining the functional formula for the sum of a series $g_n = \sum_{k=0}^{n-1} f_k$ can be reduced to the solution of the nonhomogeneous difference equation of first order

$$\Delta g_n = g_{n+1} - g_n = f_n . \tag{3.15}$$

This is easily verified by substituting successively into (3.15)

$$\begin{aligned} g_1 - g_0 &= f_0 , \\ g_2 - g_1 &= f_1 , \\ g_3 - g_2 &= f_2 , \\ \vdots \quad \vdots \\ g_n - g_{n-1} &= f_{n-1} \end{aligned}$$

and adding up all these equations

$$g_n - g_0 = \sum_{k=0}^{n-1} f_k . \tag{3.16}$$

If we are looking for the sum of an infinite series (under the assumption of the existence of the sum), we have to determine the limit of the sequence of partial sums (3.16) for $n \to +\infty$:

$$\lim_{n \to +\infty} g_n = \lim_{n \to +\infty} \sum_{k=0}^{n-1} f_k = \sum_{k=0}^{+\infty} f_k . \tag{3.17}$$

Thus, we investigate in fact the asymptotic behaviour of the sum $\sum_{k=0}^{n-1} f_k$ for $n \to +\infty$.

In this section we shall discuss the determination of the formula for the sum of finite as well as infinite series using the mathematical tools of the $\mathscr{Z}$ transform. We have to take into account that besides series for whose sum it is possible to find a formula there also exist such convergent series, e.g. $\sum_{k=1}^{n} k^{-2}$, for whose sum the formula does not exist. In such cases not even the application of the $\mathscr{Z}$ transform solves the problem since an integral is obtained which can be solved only by expansion into a series whose sum we seek to express by the formula.

The summation of finite series consists in the application of the $\mathscr{Z}$ transform to the sequence of partial sums (Theorem 19). The approach is illustrated using the example of the sequence $\{f_n\} = \{\sin^2 \beta n\}$. For the nth partial sum we have

$$g_n = \sum_{k=0}^{n-1} \sin^2 \beta k . \tag{3.18}$$

First, we determine the transform of the sequence $\{f_n\}$. If the familiar formula $\sin^2 \beta n = (1 - \cos 2\beta n)/2$ is applied, the list of transforms yields

$$\mathscr{Z}\{\sin^2 \beta n\} = \frac{1}{2}\left[\frac{z}{z-1} - \frac{z(z - \cos 2\beta)}{z^2 - 2z\cos 2\beta + 1}\right]. \tag{3.19}$$

Thus, the transform of the sequence of partial sums is

$$\mathscr{Z}\{g_n\} = \frac{1}{z-1}\,\frac{1}{2}\left[\frac{z}{z-1} - \frac{z(z - \cos 2\beta)}{z^2 - 2z\cos 2\beta + 1}\right].$$

The inverse transform is obtained by the residue theorem (2.35). The desired formula for the partial sum of the series is

$$\sum_{k=0}^{n-1} \sin^2 \beta k = \frac{1}{2}n - \frac{1}{4}\left(1 + \frac{1 - \cos 2\beta n}{1 - \cos 2\beta}\right). \tag{3.20}$$

The process of the summation of infinite series is simpler than for the derivation of the formula for the sequence of partial sums. It consists merely in finding the transform of the sequence whose infinite sum we are looking for, and in the application of Theorem 16. The procedure is illustrated with the aid of the example of the series

$$\sum_{n=0}^{+\infty} \frac{\cos \beta(2n+1)}{2n+1} \quad \text{and} \quad \sum_{n=0}^{+\infty} \frac{\sin \beta(2n+1)}{2n+1} \tag{3.21}$$

for $\quad 0 < \beta < \pi$.

Apply the Euler formula

$$e^{j\beta(2n+1)} = \cos \beta(2n+1) + j \sin \beta(2n+1)$$

and Theorem 15 on integration with respect to a parameter. Since we have

$$\frac{a^{2n+1}}{2n+1} = \int_0^a \xi^{2n}\, d\xi\,,$$

we proceed when determining the transform as follows:

$$\mathscr{Z}\left\{\frac{a^{2n+1}}{2n+1}\right\} = \mathscr{Z}\left\{\int_0^a \xi^{2n}\, d\xi\right\} = \int_0^a \mathscr{Z}\{\xi^{2n}\}\, d\xi =$$

$$= \int_0^a \frac{z}{z - \xi^2}\, d\xi = \frac{\sqrt{z}}{2} \ln \frac{\sqrt{z} + a}{\sqrt{z} - a}, \qquad |z| > |a^2|. \tag{3.22}$$

Further, if we put $a = e^{j\beta}$, relation (3.22) implies

$$\mathscr{Z}\left\{\frac{e^{j\beta(2n+1)}}{2n+1}\right\} = \frac{\sqrt{z}}{2} \ln \frac{\sqrt{z} + \cos\beta + j\sin\beta}{\sqrt{z} - \cos\beta - j\sin\beta} =$$

$$= \frac{\sqrt{z}}{2} \ln \left[\sqrt{\frac{z + 1 + 2(\sqrt{z})\cos\beta}{z + 1 - 2(\sqrt{z})\cos\beta}}\, e^{j\arctan[2(\sqrt{z})\sin\beta/(z-1)]} \right] =$$

$$= \frac{\sqrt{z}}{4} \ln \frac{z + 1 + 2(\sqrt{z})\cos\beta}{z + 1 - 2(\sqrt{z})\cos\beta} + j\frac{\sqrt{z}}{2} \arctan \frac{2(\sqrt{z})\sin\beta}{z - 1}. \tag{3.23}$$

Comparing the real and imaginary parts we obtain the transforms of the sequences whose infinite sums we are seeking:

$$\mathscr{Z}\left\{\frac{\cos\beta(2n+1)}{2n+1}\right\} = \frac{\sqrt{z}}{4} \ln \frac{z + 1 + 2(\sqrt{z})\cos\beta}{z + 1 - 2(\sqrt{z})\cos\beta}, \tag{3.24a}$$

$$\mathscr{Z}\left\{\frac{\sin\beta(2n+1)}{2n+1}\right\} = \frac{\sqrt{z}}{2} \arctan \frac{2(\sqrt{z})\sin\beta}{z - 1}. \tag{3.24b}$$

Under the assumption of the convergence of the series (3.21) – which is satisfied in our case – Theorem 16 can be applied and we obtain

$$\sum_{n=0}^{+\infty} \frac{\cos\beta(2n+1)}{2n+1} = \lim_{\substack{z\to 1+ \\ \operatorname{Im} z = 0}} \mathscr{Z}\left\{\frac{\cos\beta(2n+1)}{2n+1}\right\} =$$

$$= \frac{1}{4} \ln \frac{1 + \cos\beta}{1 - \cos\beta} = \frac{1}{4} \ln \cot^2 \frac{\beta}{2}, \tag{3.25a}$$

$$\sum_{n=0}^{+\infty} \frac{\sin\beta(2n+1)}{2n+1} = \lim_{\substack{z\to 1+ \\ \operatorname{Im} z = 0}} \mathscr{Z}\left\{\frac{\sin\beta(2n+1)}{2n+1}\right\} =$$

$$= \frac{1}{2} \arctan(+\infty) = \frac{\pi}{4}. \tag{3.25b}$$

3.1.2. *Chain of Two-Ports*

As an example of the solution of a system of difference equations we may mention the problem of the determination of the voltage and current at the output of a chain of general two-ports whose parameters vary according to a certain formula. Simultaneously, this is a problem in which not time but the order of the two-ports is the discrete variable.

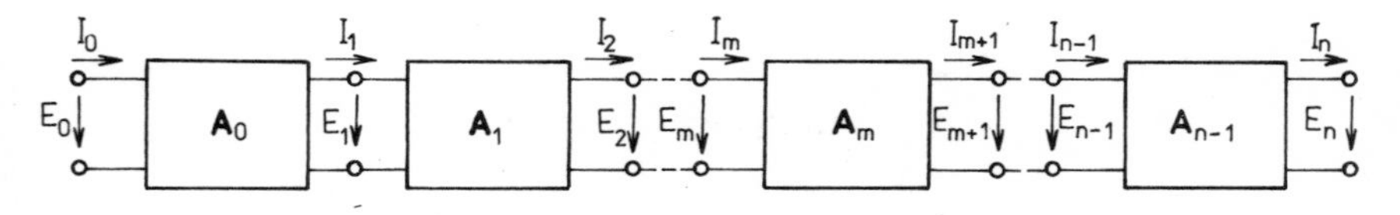

Fig. 21

Let a chain of n two-ports be given (see Fig. 21) each of which is described by the matrix equation

$$\begin{bmatrix} E_{m+1} \\ I_{m+1} \end{bmatrix} = \begin{bmatrix} a & bK^{-m} \\ cK^m & d \end{bmatrix} \begin{bmatrix} E_m \\ I_m \end{bmatrix} = \mathbf{A}_m \begin{bmatrix} E_m \\ I_m \end{bmatrix}. \tag{3.26}$$

The subscript m, $m = 0, 1, 2, \ldots, n-1$, denotes the order of the two-ports starting from the input of the chain. The functions $E_m = E_m(p)$ and $I_m = I_m(p)$ are Laplace transforms of the voltage and current on the mth terminals. The symbols a, b, c, d are the two-port parameters of the inverse transmission matrix. The coefficient $K > 0$ represents the "progression factor" of the two-ports. The value $K = 1$ corresponds to identical two-ports.

For the entire chain we obviously have

$$\begin{bmatrix} E_n \\ I_n \end{bmatrix} = \prod_{m=0}^{n-1} \mathbf{A}_m \begin{bmatrix} E_0 \\ I_0 \end{bmatrix} \quad \text{for} \quad n = 0, 1, 2, \ldots. \tag{3.27}$$

Let us find the formula for the elements of the matrix $\sum_{m=0}^{n-1} \mathbf{A}_m$ in dependence on the number of two-ports, i.e. on the subscript n.

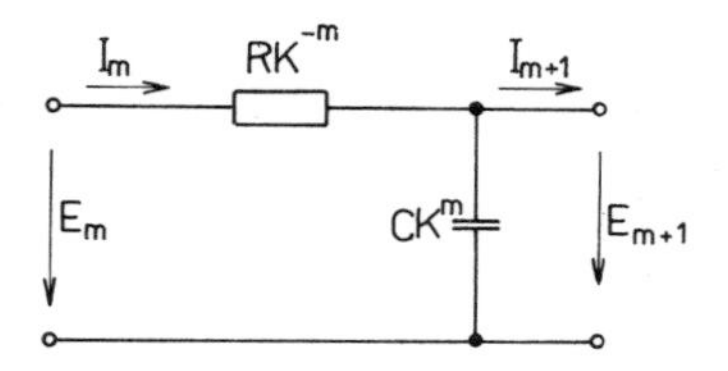

Fig. 22

A problem of this type is encountered for instance when constructing a chain of RC networks each of which is constituted by a circuit given by Fig. 22. Each network is characterized by an inverse transmission matrix

$$\mathbf{A}_m = \begin{bmatrix} 1 & RK^{-m} \\ pCK^m & 1 + pRC \end{bmatrix}.$$

This matrix equation corresponds to a system of two difference equations of first order with variable coefficients. To their solution the $\mathscr{Z}$ transform is applied. To simplify the procedure of solution we shall not directly solve equation (3.26) but the following equivalent system obtained by row and column manipulations:

$$\begin{bmatrix} K^m E_{m+1} \\ I_{m+1} \end{bmatrix} = \begin{bmatrix} a & b \\ c & d \end{bmatrix} \begin{bmatrix} K^m E_m \\ I_m \end{bmatrix} = \mathbf{A} \begin{bmatrix} K^m E_m \\ I_m \end{bmatrix}. \tag{3.28}$$

Assume that the sequences of functions $\{E_m\}$ and $\{I_m\}$ are $\mathscr{Z}$ transformable and that we have $\mathscr{Z}\{E_m\} = E(z)$, $\mathscr{Z}\{I_m\} = I(z)$. The $\mathscr{Z}$ transform of relation (3.28) and the application of the translation theorem and of the similarity theorem yield the transform of the matrix equation

$$z \begin{bmatrix} K^{-1} \left[E\left(\dfrac{z}{K}\right) - E_0 \right] \\ I(z) - I_0 \end{bmatrix} = \mathbf{A} \begin{bmatrix} E\left(\dfrac{z}{K}\right) \\ I(z) \end{bmatrix}, \tag{3.29}$$

where E_0 and I_0 are the Laplace transforms of the voltage and current at the input of the chain. Hence, we obtain

$$(K^{-1/2} z \mathbf{K}_1 - \mathbf{A}) \begin{bmatrix} E\left(\dfrac{z}{K}\right) \\ I(z) \end{bmatrix} = K^{-1/2} z \mathbf{K}_1 \begin{bmatrix} E_0 \\ I_0 \end{bmatrix}, \tag{3.30}$$

where

$$\mathbf{K}_1 = \begin{bmatrix} K^{-1/2} & 0 \\ 0 & K^{1/2} \end{bmatrix}.$$

Under the assumption of the existence of the inverse matrix to the matrix $K^{-1/2}z\boldsymbol{K}_1 - \boldsymbol{A}$, the solution of equation (3.30) yields the column matrix of transforms

$$\begin{bmatrix} E\left(\frac{z}{K}\right) \\ I(z) \end{bmatrix} = \frac{K^{1/2}z}{z^2 - K^{1/2}uz + Ky}\left(K^{-1/2}z\boldsymbol{1} - \tilde{\boldsymbol{A}}\boldsymbol{K}_1\right)\begin{bmatrix} E_0 \\ I_0 \end{bmatrix}, \tag{3.31}$$

where $u = aK^{1/2} + dK^{-1/2}$, $y = ad - bc$. The matrix $\boldsymbol{1}$ is the unit matrix

$$\boldsymbol{1} = \begin{bmatrix} 1 & 0 \\ 0 & 1 \end{bmatrix}$$

while $\tilde{\boldsymbol{A}}$ is the adjoint matrix to the matrix $\boldsymbol{A}$

$$\tilde{\boldsymbol{A}} = \begin{bmatrix} d & -b \\ -c & a \end{bmatrix}.$$

To find the object function to the matrix of transforms (3.31) we determine first the object function

$$\mathscr{Z}^{-1}\left\{\frac{z}{z^2 - zK^{1/2}u + Ky}\right\}.$$

Two different approaches will be used here:

(a) Expansion into a series is of the form

$$\begin{aligned} \frac{z}{z^2 - zK^{1/2}u + Ky} &= z^{-1} + K^{1/2}\, Q_2(u, y)\, z^{-2} + \\ &+ K\, Q_3(u, y)\, z^{-3} + K^{3/2}\, Q_4(u, y)\, z^{-4} + \ldots = \\ &= \mathscr{Z}\{K^{(n-1)/2}\, Q_n(u, y)\}\,. \end{aligned} \tag{3.32}$$

It can be shown that the functions $Q_n(u, y)$ are polynomials in the variables u and y for which we have the recurrent relation

$$Q_n(u, y) = u\, Q_{n-1}(u, y) - y\, Q_{n-2}(u, y) \tag{3.33}$$

for $n \geqq 2$, with initial terms $Q_0(u, y) = 0$ and $Q_1(u, y) = 1$.

(b) The integral of the inverse $\mathscr{Z}$ transform, or rather the list of transforms, yields the following formula for the object function:

$$\mathscr{Z}^{-1}\left\{\frac{z}{z^2 - zK^{1/2}u + Ky}\right\} =$$

$$= (Ky)^{(n-1)/2} \frac{\sin\left(n \arccos \frac{1}{2} uy^{-1/2}\right)}{\sin\left(\arccos \frac{1}{2} uy^{-1/2}\right)}. \tag{3.34}$$

Comparing the two expressions for the object functions we obtain the expression

$$Q_n(u, y) = y^{(n-1)/2} \frac{\sin\left(n \arccos \frac{1}{2} uy^{-1/2}\right)}{\sin\left(\arccos \frac{1}{2} uy^{-1/2}\right)} \tag{3.35}$$

for the polynomials $Q_n(u, y)$. Upon substitution into (3.31) and the application of the translation theorem, the transform of the solution of the system of equations is

$$\begin{bmatrix} E\left(\frac{z}{K}\right) \\ I(z) \end{bmatrix} =$$

$$= \left(\mathbf{1}\, \mathscr{Z}\{K^{n/2} Q_{n+1}(u, y)\} - K^{1/2}\tilde{\mathbf{A}}\mathbf{K}_1\, \mathscr{Z}\{K^{(n-1)/2} Q_n(u, y)\}\right) \begin{bmatrix} E_0 \\ I_0 \end{bmatrix}. \tag{3.36}$$

The inverse transform of the matrix of transforms (3.36) yields the solution of the equivalent system:

$$\begin{bmatrix} K^n E_n \\ I_n \end{bmatrix} =$$

$$= \begin{bmatrix} K^{n/2} Q_{n+1}(u, y) - dK^{(n-1)/2} Q_n(u, y), & bK^{(n+1)/2} Q_n(u, y) \\ cK^{(n-1)/2} Q_n(u, y), & K^{n/2} Q_{n+1}(u, y) - aK^{(n+1)/2} Q_n(u, y) \end{bmatrix} \begin{bmatrix} E_0 \\ I_0 \end{bmatrix} \tag{3.37}$$

In (3.37) it remains to express the polynomial $Q_{n+1}(u, y)$ by the recurrent relation. The voltage and current at the output of a chain of n variable elements are then given by the matrix equation

$$\begin{bmatrix} E_n \\ I_n \end{bmatrix} =$$
$$= \begin{bmatrix} aK^{-(n-1)/2} Q_n(u, y) - yK^{-n/2} Q_{n-1}(u, y), & bK^{-(n-1)/2} Q_n(u, y) \\ cK^{(n-1)/2} Q_n(u, y), & dK^{(n-1)/2} Q_n(u, y) - yK^{n/2} Q_{n-1}(u, y) \end{bmatrix} \begin{bmatrix} E_0 \\ I_0 \end{bmatrix}, \tag{3.38}$$

which can be written in the following more concise form:

$$\begin{bmatrix} E_n \\ I_n \end{bmatrix} = (\mathbf{K}_{n-1} \mathbf{A}\, Q_n(u, y) - y\mathbf{K}_n\, Q_{n-1}(u, y)) \begin{bmatrix} E_0 \\ I_0 \end{bmatrix} \tag{3.39}$$

for

$$\mathbf{K}_n = \begin{bmatrix} K^{-n/2} & 0 \\ 0 & K^{n/2} \end{bmatrix}, \qquad \mathbf{A} = \begin{bmatrix} a & b \\ c & d \end{bmatrix}.$$

Comparing (3.39) with (3.27) we obtain the following relation for the inverse transmission matrix of a chain of two-ports:

$$\prod_{m=0}^{n-1} \mathbf{A}_m = \mathbf{K}_{n-1} \mathbf{A}\, Q_n(u, y) - y\mathbf{K}_n\, Q_{n-1}(u, y)\,. \tag{3.40}$$

Let us mention several further cases of specific kinds of chains. If we have a chain of reciprocal two-ports, then $y = ad - bc = 1$ holds and the resulting matrix is

$$\prod_{m=0}^{n-1} \mathbf{A}_m = \mathbf{K}_{n-1} \mathbf{A}\, Q_n(u, 1) - \mathbf{K}_n\, Q_n(u, 1)\,, \tag{3.41}$$

where

$$Q_n(u, 1) = \frac{\sin\left(n \arccos \frac{1}{2} u\right)}{\sin\left(\arccos \frac{1}{2} u\right)},$$

$$Q_n(u, 1) = u\, Q_{n-1}(u, 1) - Q_{n-2}(u, 1)\,,$$
$$Q_0(u, 1) = 0\,, \qquad Q_1(u, 1) = 1\,.$$

Putting $K = 1$, i.e. if we have a chain of identical two-ports, we have $\mathbf{A}_m = \mathbf{A}_{m-1}$ for $m = 1, 2, \ldots, n-1$. Then the inverse transmission matrix of general two-ports is

$$\mathbf{A}^n = \mathbf{A}\, Q_n(x, y) - y\mathbf{1}\, Q_{n-1}(x, y), \tag{3.42}$$

where $x = a + d$, $y = ad - bc$. For a chain of reciprocal identical two-ports we again put $y = 1$.

3.2. Transfer Function of a Discrete System

In this section some important concepts from the theory of linear discrete systems will be explained which are being applied to the analysis, simulation and design of systems for discrete signal processing.

We follow the considerations performed in Section 3.1 when solving linear difference equations with constant coefficients with the aid of the $\mathscr{Z}$ transforms.

Let us start from the recurrent equation of sth order (3.3) where – taking account of the generality of the considerations below – the order r is used on the right-hand side:

$$\begin{aligned} &b_s h_{n+s} + b_{s-1} h_{n+s-1} + \ldots + b_0 h_n = \\ &= a_r f_{n+r} + a_{r-1} f_{n+r-1} + \ldots + a_0 f_n . \end{aligned}$$

Assume that the system described by this equation was at rest for $n < 0$, i.e. assume that we have $f_n = h_n = 0$ for $n < 0$. We apply the $\mathscr{Z}$ transform and obtain the transform of the particular solution of the difference equation in the form

$$H(z) = \frac{\sum_{i=0}^{r} a_i z^i}{\sum_{i=0}^{s} b_i z^i} F(z), \tag{3.43}$$

where $H(z) = \mathscr{Z}\{h_n\}$ and $F(z) = \mathscr{Z}\{f_n\}$. The ratio of the transform of the particular solution $H(z)$ and the transform $F(z)$ defines a function which completely characterizes the discrete system described by the difference equation. This function is denoted by

$$G(z) = \frac{H(z)}{F(z)} = \frac{\sum_{i=0}^{r} a_i z^i}{\sum_{i=0}^{s} b_i z^i} \tag{3.44}$$

and called the *transfer function of a linear discrete system.* Relation (3.43) can then be briefly written in the form

$$H(z) = F(z)\, G(z)\,. \tag{3.45}$$

It is represented symbolically by a block diagram, as given in Fig. 23.

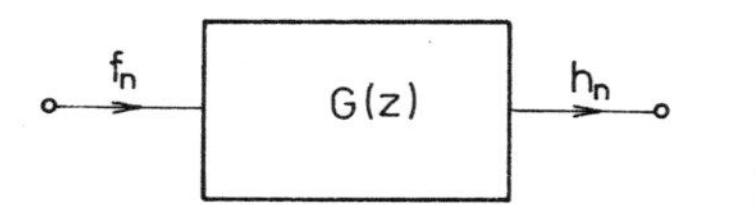

Fig. 23

The sequence $\{f_n\}$ is called the *input signal,* while the particular solution of the difference equation is called the *output signal* of the discrete system with transfer function $G(z)$.

From relation (3.45) we derive further implications for the transfer function. If we put $F(z) = 1$ here, which corresponds to the transform of a unit impulse defined by the relations

$$f_0 = \delta_0 = 1\,,$$
$$f_n = \delta_n = 0 \qquad \text{for} \qquad n \neq 0\,,$$

we obtain the equality $H(z) = G(z)$. Hence, applying the inverse transform, we have $\{h_n\} = \{g_n\}$. In other words: *The transfer function $G(z)$ is the transform of the impulse response of a discrete system.*

Now, let us return to relation (3.45) in which the transform of the response is given as the product of the transfer function $G(z)$ and the transform of the input signal $F(z)$. However, according to Theorem 9, the product of the transforms is the transform of the convolution of the sequences. Thus, the response of a discrete system $\{h_n\}$ can be written in the form

$$h_n = \sum_{k=0}^{n} f_k g_{n-k}, \tag{3.46}$$

where $\{f_n\}$ is the input signal and $\{g_n\}$ is the *impulse response of the discrete system.*

The transfer function $G(z)$ is given in relation (3.44) as the ratio of two polynomials in the variable z

$$G(z) = \frac{P(z)}{Q(z)},$$

where $P(z) = \sum_{i=0}^{r} a_i z^i$ and $Q(z) = \sum_{i=0}^{s} b_i z^i$. Frequently, it is useful to represent it with the aid of zeros and poles as

$$G(z) = K \frac{\prod_{\mu=1}^{r} (z - z_\mu)}{\prod_{\nu=1}^{s} (z - z_\nu)}. \tag{3.47}$$

Here, the zeros z_μ, $\mu = 1, 2, \ldots, r$, are obtained by solving the equation $P(z) = 0$ while the poles z_ν, $\nu = 1, 2, \ldots, s$, are obtained by solving the equation $Q(z) = 0$. For the constant K we have $K = a_r/b_s$. As far as the polynomials $P(z)$ and $Q(z)$ have real coefficients, the zeros and poles are either real or they occur in complex conjugate pairs.

So far no assumption was made concerning the orders of the numerator and the denominator. It is obvious that when using the one-sided $\mathscr{Z}$ transform it must hold that $r \leqq s$. If this assumption is not satisfied the transfer function is a two-sided transform of the impulse response. This case is related to a further important property of discrete systems, namely to causality which is discussed in Subsection 3.2.1.

Until now the transfer function $G(z)$ of a discrete system was assumed to be given by the ratio of two polynomials in the variable z. This ratio was expressed in relation (3.47) with the aid of zeros and poles. The corresponding impulse response $\{g_n\}$ is then given by a linear combination of exponential sequences (under the assumption that not all poles $z_\nu = 0$).

In this case, there does not exist a number $N > 0$ such that for $n > N$ we have $g_n = 0$. A discrete system with this property is called a *system with infinite impulse response.* However, there exist discrete systems for which the coefficients in the denominator of the transfer function are $b_i = 0$ for $i = 0, 1, \ldots, s-1$. This corresponds to the condition $z_v = 0$ for $v = 1, 2, \ldots, s$. Then the transfer function possesses a single pole of order s at the point $z = 0$ and it assumes the form

$$G(z) = \frac{\sum_{i=0}^{r} a_i z^i}{b_s z^s} = \frac{1}{b_s} \sum_{i=0}^{r} a_i z^{i-s}. \tag{3.48}$$

The corresponding impulse response $\{g_n\}$ is

$$g_n = \mathscr{Z}^{-1}\{G(z)\} = \frac{1}{b_s} \sum_{i=0}^{r} a_i \delta_{n-s+i}, \tag{3.49}$$

where δ_{n-k} is the unit impulse at the point k for which

$$\begin{aligned} \delta_{n-k} &= 1 \qquad \text{for} \qquad n = k, \\ \delta_{n-k} &= 0 \qquad \text{for} \qquad n \neq k. \end{aligned}$$

A discrete system with transfer function (3.48) is called a *system with finite impulse response.*

3.2.1. *Stability and Causality of Discrete Systems*

In the preceding subsection it was verified that a discrete transfer function $G(z) = \mathscr{Z}\{g_n\}$ characterizes exhaustively the properties of a discrete system, which is described by a difference equation with constant coefficients, under the assumption that the system is at rest for $n < 0$. Now, we shall discuss two important properties of discrete systems which specify a narrower class of transfer functions. These properties are *stability* and *causality.*

A *stable discrete system* is defined as a system whose response to every bounded input signal is also bounded. Obviously, a bounded input signal satisfies the condition $|f_n| < M < +\infty$ for all $n \geqq 0$. The response of a discrete system is given by the convolution

$$h_n = \sum_{k=0}^{n} f_k g_{n-k}. \tag{3.50}$$

Consequently, for stable systems

$$|h_n| = \left| \sum_{k=0}^{n} g_k f_{n-k} \right| \leqq M \sum_{k=0}^{+\infty} |g_k| < +\infty$$

must hold. This implies the following necessary and sufficient condition for the stability of a linear discrete system:

$$\sum_{k=0}^{+\infty} |g_k| < +\infty . \tag{3.51}$$

The sequence $\{g_n\}$ is the impulse response of the system, and we have $\{g_n\} = \mathscr{Z}^{-1}\{G(z)\}$. From the properties of the $\mathscr{Z}$ transform it follows that for the fulfilment of condition (3.51) the transform $G(z)$ must exist for $|z| \geqq 1/R = 1$. However, this means that the transfer function, which is the transform of the impulse response, is a regular function in the region $|z| \geqq 1$. Consequently, it is possible to formulate the necessary and sufficient condition for the stability of a linear discrete system with the aid of the poles of the transfer function for which the inequality

$$|z_\nu| < 1 \qquad \text{must hold for} \qquad \nu = 1, 2, \ldots, s . \tag{3.52}$$

Thus the poles of the transfer function of a linear discrete system must lie inside the unit circle. It is obvious that systems with finite impulse response whose all poles lie at the point $z = 0$ satisfy the condition and are thus always stable.

Let us discuss the concept of causality now. A *system* is said to be *causal* if the value of its response $\{h_n\}$, for arbitrary $n = n_0$, depends only on terms of the input signal $\{f_n\}$ for $n \leqq n_0$. In other words: If the input signal is $f_n \equiv 0$ for $n \leqq n_0$, then for a causal system we have $h_n \equiv 0$ for $n \leqq n_0$. This implies the following necessary and sufficient condition for the causality of a discrete system imposed on the impulse response $\{g_n\}$:

$$g_n = 0 \qquad \text{for} \qquad n < 0 . \tag{3.53}$$

Further, we shall investigate how the causality condition (3.53) manifests itself on the form of the transform function $G(z)$. From the properties of the $\mathscr{Z}$ transform it follows that the transform of a sequence which satisfies (3.53) is a regular function for $|z| > 1/R$. Such a function is regular at the point $z = \infty$, and thus – as far as it has the form of a rational function – the polynomial in the numerator has to be at most of the same degree as the polynomial in the denominator. Thus, if a linear discrete system is to be causal, then for its transfer function $G(z)$ in the form (3.43) we must have $r \leqq s$. Consequently, a causal system is defined by its transfer function which is the transform of the impulse response in the one-sided $\mathscr{Z}$ transform.

If the causality condition (3.53) is not satisfied, then the transfer function of a discrete system is the transform of the impulse response in the sense of the definition of the two-sided $\mathscr{Z}$ transform, and we can denote it by $G_{\text{II}}(z)$. Then the *discrete system* is *non-causal*, and for its transfer function $G_{\text{II}}(z)$ we have $r > s$. A transfer function of this type can be decomposed into two partial one-sided transfer functions in the sense of Section 2.4:

$$G_{\text{II}}(z) = G_+(z) + G_-(z) \tag{3.54}$$

where

$$G_+(z) = \sum_{n=0}^{+\infty} g_n z^{-n} \qquad \text{and} \qquad G_-(z) = \sum_{n=-1}^{-\infty} g_n z^{-n}.$$

Obviously, the function $G_+(z)$ satisfies condition (3.53), thus it is the *causal part* of the two-sided transfer function $G_{\text{II}}(z)$. A function $G_-(z)$ which does not satisfy the causality condition is called the *anticipative part* of the two-sided transfer function. The response $\{h_n\}$ of a non-causal system to an input signal $\{f_n\}$ does also not satisfy the condition $h_n = 0$ for $n < 0$.

Furthermore, we discuss the connection between stability and causality of a discrete system. If we define the transfer function in the sense of the one-sided $\mathscr{Z}$ transform, then the discrete system will always be a causal system and it may be either stable or unstable. However, the transfer function of a non-causal discrete system must be understood in the sense of a two-sided $\mathscr{Z}$ transform, and then not only the position of the poles of the transfer function but also the choice of the regularity region of the function $G(z)$ determine its stability. This means that an unstable causal system, which is described in the one-sided $\mathscr{Z}$ transform by a transfer function regular in the region $|z| > 1/R > 1$, may be interpreted as a non-causal stable system in the two-sided $\mathscr{Z}$ transform with a transfer function which is regular in the annulus $1/R_+ \leqq 1 \leqq |z| < R_-$, where $R_- = 1/R$. These assertions can be easily derived from the considerations of the examples of Section 2.4.

EXAMPLE 1. Let the transfer function of a discrete system be given by the function

$$G(z) = \frac{z}{(z-a)(z-b)},$$

regular in the region $|z| > \max(|a|, |b|)$. The system is causal since the degree of the numerator is smaller than the degree of the denominator. The function $G(z)$ is the transform of the impulse response

$$g_n = \mathscr{Z}^{-1}\{G(z)\} = \frac{1}{a-b}(a^n - b^n)$$

for $n \geqq 0$. If both poles $z = a$ and $z = b$ lie inside the unit circle, the system is stable. Moreover, should any of the poles lie outside the unit circle, the causality of the system would not change but the system would be unstable.

EXAMPLE 2. Now, let us assume that $|a| > 1$, $|b| < 1$ holds in the transfer function of the previous example. We denote this function by $G_{II}(z)$ and, in the sense of the two-sided $\mathscr{Z}$ transform, understand it to be a function regular in the annulus $b < |z| < a$, which is the transform of the sequence $\{g_n\}$ and for which we have (see Subsection 2.4.1)

$$g_n = -\frac{b^n}{a-b} \quad \text{for} \quad n \geqq 0,$$

$$g_n = -\frac{a^n}{a-b} \quad \text{for} \quad n < 0.$$

The sequence $\{g_n\}$ does not satisfy condition (3.53) and, therefore, the discrete system with transfer function $G_{II}(z)$ is not causal. Let us investigate its stability now. A part of the sequence $\{g_n\}$ for $n \geqq 0$ satisfies the condition of stability (3.51) since $|b| < 1$. For the left-hand part of the sequence $\{g_n\}$, i.e., for $n < 0$, we have

$$\sum_{n=-1}^{-\infty} |g_n| = \sum_{n=1}^{+\infty} |g_{-n}| = \frac{1}{|a-b|} \sum_{n=1}^{+\infty} \left|\frac{1}{a}\right|^n < +\infty .$$

Thus, the condition of stability is satisfied for the left-hand part of the sequence $\{g_n\}$ as well.

EXAMPLE 3. Let us investigate the properties of a discrete system whose transfer function has the poles $z = a$ and $z = b$ which lie inside the unit circle but whose numerator is of higher degree than the denominator. Obviously, such a transfer function is a regular function in the annulus $\max(|a|, |b|) < 1 < |z| < +\infty$ and it has to be understood in the sense of the two-sided $\mathscr{Z}$ transform. Let this transfer function be of the form

$$G_{II}(z) = \frac{z^3}{(z - a)(z - b)}.$$

The function is decomposed upon division of the numerator by the denominator into a left and right transform as follows

$$G(z) = z + \frac{z[z(a + b) - ab]}{(z - a)(z - b)}.$$

Hence, upon application of the inverse two-sided $\mathscr{Z}$ transform, we obtain the sequence $\{g_n\}$ given by the formula

$$\begin{aligned} g_n &= \frac{1}{a - b}(a^{n+2} - b^{n+2}) \quad &\text{for} \quad n \geqq 0, \\ g_{-1} &= 1, \\ g_n &= 0 \quad &\text{for} \quad n < -1. \end{aligned}$$

Obviously, the result indicates that the discrete system is non-causal but stable.

3.2.2. *Frequency Characteristics of Linear Discrete Systems*

The frequency characteristics are important elements of the analysis and synthesis of discrete systems since they facilitate the assessment of the systems' properties. We shall show the connection of the frequency properties of discrete systems with transfer functions.

The response of a linear discrete system to an input signal is given by the convolution

$$h_n = \sum_{k=0}^{n} f_{n-k} g_k.$$

As the input signal let us use the complex sequence $\{f_n\} = \{e^{j\omega n}\} = \{\cos \omega n + j \sin \omega n\}$ defined for $-\infty < n < +\infty$. If the system is causal, for its response we have

$$h_n = \sum_{k=0}^{+\infty} e^{j\omega(n-k)} g_k = e^{j\omega n} \sum_{k=0}^{+\infty} g_k \, e^{-j\omega k}. \tag{3.55a}$$

On the right-hand side of the relation we recognize the definitorical relation of the $\mathscr{Z}$ transform in which $z = e^{j\omega}$. Thus, relation (3.55a) can be written in the form

$$h_n = e^{j\omega n} \, G(e^{j\omega}). \tag{3.55b}$$

Consequently, the response $\{h_n\}$ of a discrete system to the input sequence $\{f_n\} = \{e^{j\omega n}\}$ is given by its product with the value of the transfer function $G(z)$ for $z = e^{j\omega}$ which lies on the unit circle, with ω being the radian frequency of the input signal. The function $G(e^{j\omega})$ is called the *complex frequency response* of a linear discrete system. It can be represented in the form

$$G(e^{j\omega}) = \operatorname{Re} G(e^{j\omega}) + j \operatorname{Im} G(e^{j\omega}) = A(\omega) \, e^{j\varphi(\omega)}, \tag{3.56}$$

where $A(\omega) = |G(e^{j\omega})|$ is the *amplitude response* while $\varphi(\omega) = \arg G(e^{j\omega})$ is the *phase response* of the transfer function of the linear discrete system. For some applications it is more adequate to use the derivative with respect to ω instead of the phase response $\varphi(\omega)$. The derivative is called the *group delay*

$$\tau(\omega) = -\frac{d\varphi(\omega)}{d\omega}.$$

The function $\tau(\omega)$ can be represented with the aid of the logarithmic derivative of the transfer function

$$\tau(\omega) = -\operatorname{Re}\left[z \frac{d}{dz} \ln G(z) \right]\Bigg|_{z=e^{j\omega}} = -\operatorname{Re}\left[z \frac{\dfrac{dG(z)}{dz}}{G(z)} \right]\Bigg|_{z=e^{j\omega}}. \tag{3.57}$$

This is called the *group delay characteristic* of a linear discrete system.

Since the exponential function is periodic with period $2\pi j$, it is obvious that

$$G(e^{j\omega}) = G(e^{j(\omega + 2k\pi)}) \tag{3.58}$$

holds for $k = \pm 1, \pm 2, \dots$. This implies that the *frequency characteristics of discrete systems are periodic with period* 2π.

On the grounds of the periodicity of the complex frequency characteristics of discrete systems it is possible to express the relationship of the $\mathscr{Z}$ transform with the Fourier transform of a sequence. Let us have the impulse response $\{g_n\}$ of a non-causal discrete system defined for $-\infty <$ $< n < +\infty$. Its transfer function is given by the two-sided $\mathscr{Z}$ transform

$$G_{\text{II}}(z) = \sum_{n=-\infty}^{+\infty} g_n z^{-n} \tag{3.59}$$

under the assumption that the series is convergent. Further, let us assume that the series is convergent in the annulus $1 - \varepsilon < |z| < 1 + \varepsilon$, where $\varepsilon > 0$ is an arbitrary small number. By Theorem 22 we then have, for the sequence $\{g_n\}$,

$$g_n = \frac{1}{2\pi\mathrm{j}} \oint_C G_{\text{II}}(z)\, z^{n-1}\, \mathrm{d}z \qquad \text{for} \qquad n = 0, \pm 1, -2, \dots, \tag{3.60}$$

where it is possible to integrate along the unit circle $|z| = 1$. Let us put $z = \mathrm{e}^{\mathrm{j}\omega}$ in (3.59) and (3.60). We obtain the pair of relations

$$G_{\text{II}}(\mathrm{e}^{\mathrm{j}\omega}) = \sum_{n=-\infty}^{+\infty} g_n\, \mathrm{e}^{-\mathrm{j}\omega n}, \tag{3.61a}$$

$$g_n = \frac{1}{2\pi} \int_0^{2\pi} G_{\text{II}}^*(\mathrm{e}^{\mathrm{j}\omega})\, \mathrm{e}^{\mathrm{j}\omega n}\, \mathrm{d}\omega. \tag{3.61b}$$

These relations are familiar from the Fourier analysis of periodic functions in the complex form. With the aid of relation (3.61b) it is possible to expand the periodic complex frequency response of a discrete system into a series in which the Fourier coefficients are equal to the terms of the sequence $\{g_n\}$. This is a surprising result, at first sight, since Fourier analysis is usually applied to the decomposition of periodic time characteristics, where the Fourier coefficients stand for the amplitudes of complex exponential functions.

The pair of relations (3.61a) and (3.61b) may be called the Fourier transform for the sequence $\{g_n\}$. We consider relation (3.61a) to be the definitorical relation of the Fourier transform (as far as the series is convergent), and relation (3.61b) to be the inverse Fourier transform. The function $G_{\text{II}}(\mathrm{e}^{\mathrm{j}\omega})$ is called the *spectrum of the sequence* $\{g_n\}$.

In concluding the considerations concerning the frequency properties of discrete systems let us still discuss the relationship of the autocorrelation sequence with the power spectral density and the complex frequency response.

In Subsection 2.3.1, Theorem 21 was employed to derive the transform of the autocorrelation sequence. On the grounds of relation (2.53a) it is possible, for the autocorrelation sequence corresponding to the impulse response $\{g_n\}$ of a discrete system, to write

$$\psi_n = \sum_{k=0}^{+\infty} g_k g_{k-n} = \frac{1}{2\pi\mathrm{j}} \oint_C G(\zeta)\, G\left(\frac{1}{\zeta}\right) \zeta^{n-1}\, \mathrm{d}\zeta\,, \tag{3.62}$$

where we integrate along the unit circle. The function $G(z)$ is the transfer function of the discrete system.

The transform of the autocorrelation sequence is given by relation (2.54)

$$\Psi(z) = \mathscr{Z}\{\psi_n\} = G(z) G\left(\frac{1}{z}\right). \tag{3.63}$$

The substitution $z = \mathrm{e}^{\mathrm{j}\omega}$ yields the spectrum of the autocorrelation sequence

$$\begin{aligned} \Psi(\mathrm{e}^{\mathrm{j}\omega}) &= \sum_{n=0}^{+\infty} \psi_n\, \mathrm{e}^{-\mathrm{j}\omega n} = G(\mathrm{e}^{\mathrm{j}\omega})\, G(\mathrm{e}^{-\mathrm{j}\omega}) = \\ &= G(\mathrm{e}^{\mathrm{j}\omega})\, G^*(\mathrm{e}^{\mathrm{j}\omega}) = |G(\mathrm{e}^{\mathrm{j}\omega})|^2\,, \end{aligned} \tag{3.64}$$

which is a real nonnegative function of the radian frequency ω equal to the square of the magnitude of the complex frequency response. The function $\Psi(\mathrm{e}^{\mathrm{j}\omega})$ is called the *power spectral density.*

Substituting $\zeta = \mathrm{e}^{\mathrm{j}\omega}$ into relation (3.62), we obtain the representation of the autocorrelation sequence by the real integral

$$\begin{aligned} \psi_n &= \frac{1}{2\pi} \int_0^{2\pi} |G(\mathrm{e}^{\mathrm{j}\omega})|^2\, \mathrm{e}^{\mathrm{j}n\omega}\, \mathrm{d}\omega = \\ &= \frac{1}{2\pi} \int_0^{2\pi} \Psi(\mathrm{e}^{\mathrm{j}\omega})\, \mathrm{e}^{\mathrm{j}n\omega}\, \mathrm{d}\omega\,, \end{aligned} \tag{3.65}$$

which is the inverse Fourier transform of the power spectral density.

The pair of equations

$$\Psi(e^{j\omega}) = \sum_{n=0}^{+\infty} \psi_n e^{-j\omega n},$$

$$\psi_n = \frac{1}{2\pi} \int_0^{2\pi} \Psi(e^{j\omega}) e^{j\omega n} d\omega, \tag{3.66}$$

which represents the relationship between the autocorrelation sequence $\{\psi_n\}$ and the power spectral density $\Psi(e^{j\omega})$ of the impulse response of a discrete system, is called the *Wiener-Khintchine theorem.*

If we put $n = 0$ in relations (3.62) and (3.65), we obtain the following expression for energy included in the impulse response of a discrete system:

$$\psi_0 = \sum_{k=0}^{+\infty} g_k^2 = \frac{1}{2\pi} \int_0^{2\pi} |G(e^{j\omega})|^2 d\omega. \tag{3.67}$$

This equality is the representation of the *Parseval theorem* for discrete systems.

3.3. Application of the $\mathscr{Z}$ Transform to the Analysis of Impulse Systems

A linear continuous system is given with the transfer function (see Fig. 24)

$$G_L(p) = \frac{U_2(p)}{U_1(p)}. \tag{3.68}$$

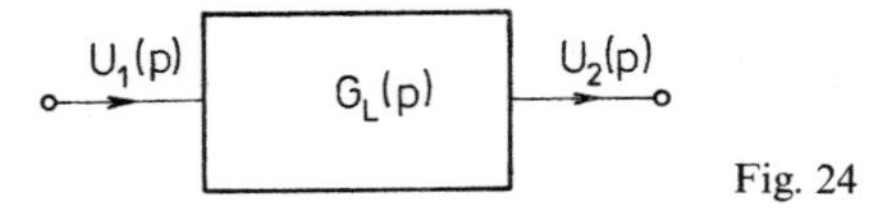

Fig. 24

Let us determine its response to a sequence of impulses of variable height. A signal of such a form is illustrated in Fig. 25 for rectangular impulses. Impulses $d(t)$ of constant width τ and unit height are modulated in their amplitude by a function $f(t)$ so that the height of the impulses is given by the values of the function $f(t)$ for $t = nT$, $n = 0, 1, 2, \dots$, $T > 0$. Obviously, we have $\tau = \text{const.}$, $0 < \tau \leqq T$. It is possible to represent the sequence of impulses by the series

$$f_1(t) = \sum_{k=0}^{[t/T]} f(kT)\, d(t - kT), \tag{3.69}$$

where $[t/T] = n$ denotes the integer part of the number t/T, i.e. the number n for which $t/T - 1 < n \leqq t/T$.

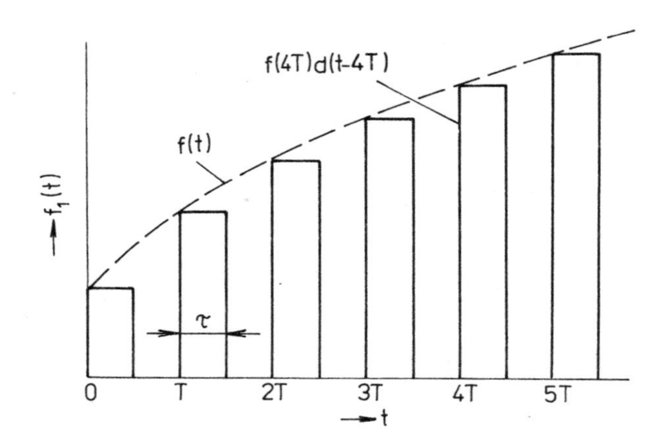

Fig. 25

In general, we distinguish two cases:

(a) The transient induced by one impulse dies out before the next impulse arrives. If we denote the response of the system to one impulse by $h(t)$, then we assume that $h(t) \doteq 0$ for $t \geqq T$. As a rule, this case is solved with the aid of the Laplace transform.

(b) The response of the system to one impulse for $t \geqq T$ did not die out yet. At time t, in general, $[t/T]$ partial responses to previous impulses are dying out and the complete response is given by their superposition. This case can be approximately solved by the summation of a bounded number of partial responses to the individual impulses. However, with the aid of the $\mathscr{Z}$ transform it is possible to solve this problem exactly and rather simply, as will be shown later.

In analogy to expression (3.69) for the input sequence of impulses, it is possible to represent the response of the system to the input sequence by the series

$$f_2(t) = \sum_{k=0}^{[t/T]} f(kT)\, h(t - kT). \tag{3.70}$$

In this expression $f(kT)\, h(t - kT)$ is the response to the kth impulse. If we limit ourselves to the points $t = nT$ when computing the response $f_2(t)$, we obtain – instead of a sum of terms of a sequence of functions – a sum of terms of a numerical sequence which is equal to the convolution of two sequences. To simplify the notation, we introduce the symbols $f_{2,n} = f_2(nT)$, $f_n = f(nT)$, $h_n = h(nT)$. For the sequence of values of the response $\{f_{2,n}\}$ of a continuous system we then have

$$f_{2,n} = \sum_{k=0}^{n} f_k h_{n-k}. \tag{3.71}$$

Upon application of the $\mathscr{Z}$ transform and of the theorem on the transform of a convolution, we obtain the transform of the sequence $\{f_{2,n}\}$

$$F_2(z) = F(z)\, H(z), \tag{3.72}$$

where $F_2(z) = \mathscr{Z}\{f_{2,n}\}$, $F(z) = \mathscr{Z}\{f_n\}$, and $H(z) = \mathscr{Z}\{h_n\}$. The transform of the sequence of values of the response of a system to an impulse of unit magnitude will be called the *impulse transfer function* $H(z)$.

Let us still recur to relation (3.69) which describes the input impulse sequence. If we put $t = nT$, and if we introduce the notation

$$f_{1,n} = f_1(nT), \qquad d_n = d(nT),$$

we obtain

$$f_{1,n} = \sum_{k=0}^{n} f_k d_{n-k}. \tag{3.73}$$

Since $\{d_n\}$ is, in consequence of $\tau \leqq T$, a sequence with a single nonzero term at the point $n = 0$ for which we have

$$d_0 = 1,$$
$$d_n = 0 \quad \text{for} \quad n \neq 0,$$

the sequence $\{d_n\}$ is identical with the unit impulse sequence $\{\delta_n\}$ whose transform is

$$D(z) = \mathscr{Z}\{\delta_n\} = 1.$$

Then the transform of the input sequence is equal to the transform of the sampled modulation function $f(t)$

$$\mathscr{Z}\{f_{1,n}\} = F_1(z) = F(z). \tag{3.74}$$

The value of the response of a system in steady state is given by the limit

$$\lim_{n\to+\infty} f_{2,n} = \lim_{n\to+\infty} \sum_{k=0}^{n} f_k h_{n-k}.$$

Let us assume that this value exists. As a rule, the correctness of this assumption can be verified by way of physical considerations. Applying Theorem 18 we then obtain

$$\lim_{n\to+\infty} f_{2,n} = \lim_{\substack{z\to 1+\\ \operatorname{Im} z = 0}} F(z)\,H(z) \tag{3.75}$$

Frequently, we need to know the values of the response of a system even at the points $t \neq nT$. Then it is possible to proceed as follows:

(a) If we wish to determine the response for all t, we introduce a new continuous variable ε in the interval $0 \leqq \varepsilon < 1$ and put $t = nT + \varepsilon T$, $n = 0, 1, 2, \ldots$. For the sequence of response values we then have, by (3.70),

$$f_2(nT + \varepsilon T) = \sum_{k=0}^{n} f(kt)\,h(nT + \varepsilon T - kT). \tag{3.76}$$

Further, we shall not deal with numerical sequences but with sequences of functions. Let us introduce the notation

$$f_{2,n}(\varepsilon) = f_2(nT + \varepsilon T), \qquad h_n(\varepsilon) = h(nT + \varepsilon T).$$

Upon applying the $\mathscr{Z}$ transform (as a rule, the conditions of transformability are not violated) we then have

$$F_2(z, \varepsilon) = F(z)\,H(z, \varepsilon), \tag{3.77}$$

where

$$F_2(z, \varepsilon) = \mathscr{Z}\{f_{2,n}(\varepsilon)\}, \qquad H(z, \varepsilon) = \mathscr{Z}\{h_n(\varepsilon)\}.$$

(b) If we are content with the values of the response of a system at equidistant points, we put $t = mT'$, where $T' = T/i$, $i > 1$ integer, $m = 0, 1, 2, \ldots$. Then we have

$$h_m = h(mT').$$

The impulse period remains without change, and $f_n = f(nT)$ holds even hereafter. Thus, the sequence $\{f_n\}$ is a decimated sequence from the sequence $\hat{f}_m = f(mT')$ according to the formula

$$\begin{aligned} \hat{f}_m &= f_n \quad \text{for} \quad m = ni\,, \\ \hat{f}_m &= 0 \quad \text{for} \quad m \neq ni\,. \end{aligned}$$

By Theorem 7a, this new sequence is also $\mathscr{Z}$ transformable and we have

$$\mathscr{Z}\{\hat{f}_m\} = F(z^i)\,.$$

For the transform of the response at the points $t = mT'$ we then have

$$F_2(z) = F(z^i)\,H(z)\,, \tag{3.78}$$

where

$$\begin{aligned} H(z) &= \mathscr{Z}\{h(mT')\}\,, \\ F(z^i) &= \mathscr{Z}\{f(nT)\}\,, \\ F_2(z) &= \mathscr{Z}\{f_2(mT')\}\,. \end{aligned}$$

3.3.1. *Response of a System to a Sequence of Rectangular Impulses*

In this subsection we apply the just performed general considerations to the solution of the case which frequently arises in practice when the input signal is constituted by a sequence of rectangular impulses.

The transform of the response to a sequence of impulses is given by relation (3.72),

$$F_2(z) = F(z)\,H(z)\,.$$

The transform $F(z)$ of the input sequence $\{f_n\}$ is determined in the usual manner. If the sequence $\{f_n\}$ is given only by a finite number of terms, it is possible to work with its transform as with a polynomial in the variable z^{-1}.

Now, we derive the expression for the impulse transfer function $H(z)$, which is the transform of the sequence of values of the response of a system to one single impulse. If we denote by $d(t)$ the rectangular impulse of width $\tau = \gamma T$, $0 < \gamma < 1$, and of unit height, then it is possible to represent it as the difference of step functions

$$d(t) = 1(t) - 1(t - \tau),$$

where

$$1(t) = 0 \quad \text{for} \quad t < 0,$$
$$1(t) = 1 \quad \text{for} \quad t \geqq 0.$$

Similarly, for the response of the system to one rectangular impulse we have

$$h(t) = a(t) - a(t - \tau),$$

where $a(t)$ is the response of the system to the unit step $1(t)$. We put $t = nT$ and introduce the notation $a_n = a(nT)$. Then the impulse transfer function is

$$\mathscr{Z}\{h_n\} = H(z) = \mathscr{Z}\{a_n\} - \mathscr{Z}\{a_n(-\gamma)\}, \tag{3.79a}$$

where $a_n(-\gamma) = a(nT - \gamma T)$. Since $a_0(-\gamma) = 0$, we prefer to write the impulse transfer function in the form

$$H(z) = a_0 + \sum_{n=1}^{+\infty} [a_n - a_n(-\gamma)]\, z^{-n}. \tag{3.79b}$$

Now, let us assume that the transfer function of the system is a rational function

$$G_L(p) = \frac{P(p)}{Q(p)}.$$

The function $P(p)$ is a polynomial of degree r while the function $Q(p)$ is a polynomial of degree s, where $r = s - 1$. We assume that the transfer function $G_L(p)$ has only simple poles, none of which lies at the origin. Then the response of the system to the unit step has the form

$$a(t) = A_0 + \sum_{i=1}^{s} A_i\, e^{p_i t}, \tag{3.80}$$

where

$$A_0 = \frac{P(0)}{Q(0)}, \qquad A_i = \frac{P(p_i)}{p_i \left[\dfrac{dQ(p)}{dp}\right]_{p = p_i}}.$$

The values p_i are roots of the equation $Q(p) = 0$. We put $t = nT$ in relation (3.80), substitute the result into (3.79a), and obtain the impulse transfer function

$$H(z) = A_0 + \sum_{i=1}^{s} A_i\left[1 + \left(1 - e^{-p_i T\gamma}\right) \sum_{n=1}^{+\infty} e^{p_i T n} z^{-n}\right],$$

where we have

$$\sum_{i=1}^{+\infty} e^{p_i T n} z^{-n} = \frac{e^{p_i T}}{z - e^{p_i T}}$$

for the partial transforms for $|z| > |e^{p_i T}|$. The resulting impulse transfer function is

$$H(z) = A_0 + \sum_{i=1}^{s} A_i \frac{z - e^{p_i T(1-\gamma)}}{z - e^{p_i T}}. \tag{3.81}$$

If some poles occur in complex conjugate pairs, the relation for the impulse transfer function can be modified so that it contains real coefficients only.

EXAMPLE. A simple RC two-port is given with the transfer function

$$G_L(p) = \frac{1}{RC} \frac{1}{p + \dfrac{1}{RC}}.$$

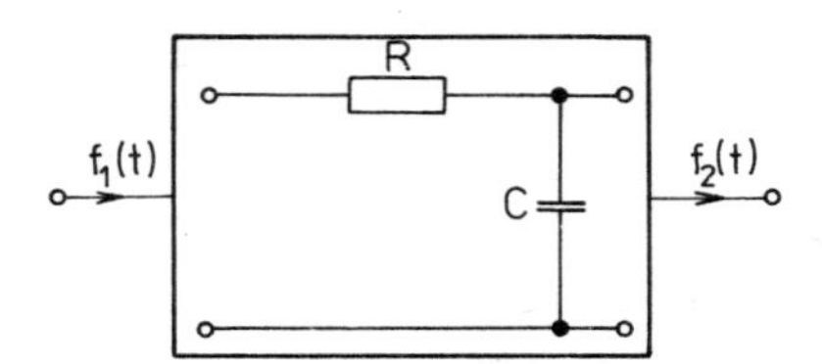

Fig. 26

The two-port is indicated in Fig. 26. A sequence of rectangular impulses of constant magnitude is fed on its input (see Fig. 27). Let us use numerical

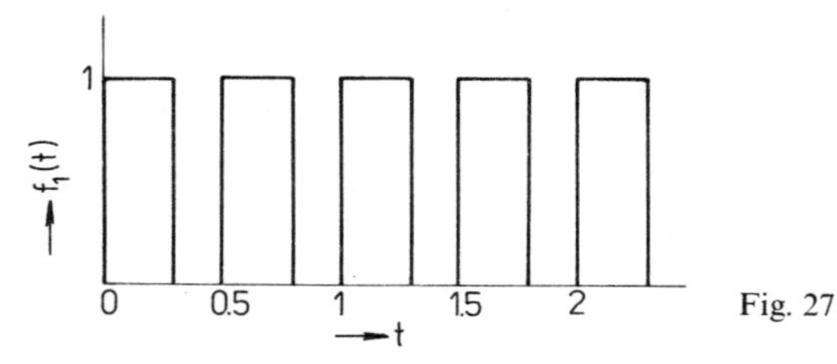

Fig. 27

values: $R = C = 1$, impulse period $T = 0.5$, impulse width $\tau = 0.3$. The relative impulse width is then $\gamma = \tau/T = 0.6$.

Upon substitution of numerical values the transfer function becomes

$$G_L(p) = \frac{1}{p + 1}.$$

For the response of the two-port to the unit step we obtain

$$a(t) = 1 - e^{-t}.$$

Then the impulse transfer function acquires the form

$$H(z) = 1 - \frac{z - e^{-0.2}}{z - e^{-0.5}} = \frac{0.212\,200}{z - 0.606\,531}.$$

The input signal is constituted by a sequence of rectangular impulses of constant height so that the input sequence $\{f_n\}$ is identical with the discrete unit sequence

$$F(z) = \frac{z}{z - 1}.$$

By (3.72) the transform of the output sequence is

$$F_2(z) = \frac{z}{z - 1}\,\frac{0.212\,200}{z - 0.606\,531} = \frac{0.212\,200\,z}{z^2 - 1.606\,531\,z + 0.606\,531}.$$

The inverse transform can be executed with the aid of Theorems 19 and 20. We apply the algebraic approach to the expansion into a series

Numerator	$\{f_{2,n}\}$
0	$0 = f_{2,0}$
0.212 200	$0.212\,200 = f_{2,1}$
	$0.340\,906 = f_{2,2}$
	$0.418\,970 = f_{2,3}$
	$0.466\,318 = f_{2,4}$
	$0.495\,036 = f_{2,5}$
	$0.512\,454 = f_{2,6}$
	$0.523\,018 = f_{2,7}$
	$\vdots \quad \vdots$

Denominator
−0.606 531
1.606 531
⟵

The value of the response in steady state is obtained by substitution into relation (3.75)

$$\lim_{n \to +\infty} f_{2,n} = \lim_{\substack{z \to 1+ \\ \operatorname{Im} z = 0}} (z - 1) F_2(z) =$$

$$= \lim_{\substack{z \to 1+ \\ \operatorname{Im} z = 0}} \frac{0.212\,200}{z - 0.606\,531} = 0.539\,305 .$$

The behaviour of the response $\{f_{2,n}\}$ to the sequence of rectangular impulses is indicated in Fig. 28 by circles.

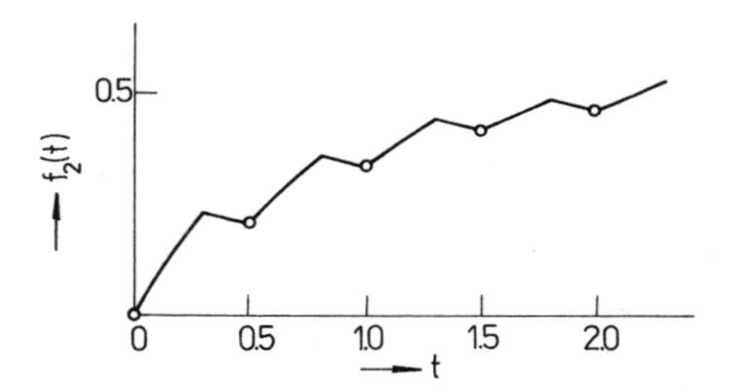

Fig. 28

3.3.2. *Evaluation of the Response of a System at an Arbitrary Point*

We take up the problem solved in the preceding subsection, but we will determine the response of the system to a sequence of rectangular impulses at the points $t = nT + \varepsilon T$ in accordance with (3.76). In this case, the transform of the output sequence is given by relation (3.77),

$$F_2(z, \varepsilon) = F(z)\, H(z, \varepsilon) .$$

First, we derive the impulse transfer function $H(z, \varepsilon)$ from the response of the system to a single impulse. This response is

$$h(t) = a(t) - a(t - \tau) ,$$

where $a(t)$ is the response of the system to the unit step. We put $t = nT + \varepsilon T$, $\tau = \gamma T$, and introduce the notation

$$h(nT + \varepsilon T) = h_n(\varepsilon) , \qquad a(nT + \varepsilon T) = a_n(\varepsilon) .$$

Then the impulse transfer function is a function of the variable z and ε,

$$\mathscr{Z}\{h_n(\varepsilon)\} = H(z, \varepsilon) = \mathscr{Z}\{a_n(\varepsilon)\} - \mathscr{Z}\{a_n(\varepsilon - \gamma)\},$$

where $a_n(\varepsilon - \gamma) = a(nT + \varepsilon T - \gamma T)$. Since $a_0(\varepsilon - \gamma) = 0$ for $\varepsilon < \gamma$, we introduce the following two relations for the impulse transfer function:

$$H_A(z, \varepsilon) = a_0(\varepsilon) + \sum_{n=1}^{+\infty} [a_n(\varepsilon) - a_n(\varepsilon - \gamma)]\, z^{-n} \qquad \text{for} \quad 0 \leqq \varepsilon < \gamma \tag{3.82a}$$

and

$$H_B(z, \varepsilon) = \sum_{n=0}^{+\infty} [a_n(\varepsilon) - a_n(\varepsilon - \gamma)]\, z^{-n} \qquad \text{for} \quad \gamma \leqq \varepsilon < 1\,. \tag{3.82b}$$

It is possible to represent the function $H_B(z, \varepsilon)$ with the aid of $H_A(z, \varepsilon)$ as follows:

$$\begin{aligned} H_B(z, \varepsilon) &= a_0(\varepsilon) - a_0(\varepsilon - \gamma) + \sum_{n=1}^{+\infty} [a_n(\varepsilon) - a_n(\varepsilon - \gamma)]\, z^{-n} = \\ &= H_A(z, \varepsilon) - a_0(\varepsilon - \gamma)\,. \end{aligned} \tag{3.83}$$

For $\varepsilon = 0$, the function $H_B(z, \varepsilon)$ has no sense while $H_A(z, \varepsilon) = H(z)$. This may be verified by the comparison of relations (3.79b) and (3.82a).

Now, we determine the impulse transfer function for the system discussed in Subsection 3.3.1 whose transfer function has the form of a rational function $G_L(p) = P(p)/Q(p)$ with simple poles none of which lies at the origin.

For the response $a(t)$ of the system to the unit step we have the expression (3.80). Upon substitution of $t = nT + \varepsilon T$ into the relation for $a(t)$, and with the aid of (3.82a) and (3.82b), we obtain the resulting impulse transfer functions

$$H_A(z, \varepsilon) = A_0 + \sum_{i=1}^{s} A_i \frac{z - e^{p_i T(1-\gamma)}}{z - e^{p_i T}}\, e^{p_i T \varepsilon} \qquad \text{for} \quad 0 \leqq \varepsilon < \gamma \tag{3.84a}$$

and

$$H_B(z, \varepsilon) = \sum_{i=1}^{s} A_i \frac{(1 - e^{-p_i T \gamma})}{z - e^{p_i T}}\, e^{p_i T \varepsilon} \qquad \text{for} \quad \gamma \leqq \varepsilon < 1\,. \tag{3.84b}$$

If some poles occur in complex conjugate pairs, it is possible to modify the partial impulse transfer functions so that they include only real coefficients.

When deriving impulse transfer functions we have assumed until now a general duration of the impulses $0 < \tau \leqq T$, where $\tau = \gamma T$, $0 < \gamma \leqq 1$. Now, we pay attention to two limit cases, namely to $\gamma = 1$ and $\gamma \to 0$. It seems reasonable to perform our considerations on the impulse transfer functions $H_A(z, \varepsilon)$ and $H_B(z, \varepsilon)$, since they allow evaluation at any arbitrary point.

For $\gamma = 1$, it suffices to deal only with the partial impulse transfer function $H_A(z, \varepsilon)$, for all $0 \leqq \varepsilon < 1$. For $\gamma = 1$, relation (3.84a) yields

$$H_1(z, \varepsilon) = A_0 + \sum_{i=1}^{s} A_i \frac{z - 1}{z - \mathrm{e}^{p_i T}} \mathrm{e}^{p_i T \varepsilon} . \tag{3.85}$$

For $\gamma \to 0$, it suffices obviously to investigate only the partial impulse transfer function $H_B(z, \varepsilon)$, for all $0 < \gamma \leqq \varepsilon < 1$. Exploiting the first two terms of the expansion of the function $\mathrm{e}^{-p_i T \gamma}$ into a series, relation (3.84b) yields the following approximate representation:

$$H_0(z, \varepsilon) \doteq \sum_{i=1}^{s} A_i \frac{p_i T \gamma z}{z - \mathrm{e}^{p_i T}} \mathrm{e}^{p_i T \varepsilon} .$$

However, since

$$A_i = \frac{P(p_i)}{p_i \left[\frac{\mathrm{d}Q(p)}{\mathrm{d}p}\right]\Bigg|_{p = p_i}} = \frac{A_i'}{p_i},$$

where

$$A_i' = \frac{P(p_i)}{\left[\frac{\mathrm{d}Q(p)}{\mathrm{d}p}\right]\Bigg|_{p = p_i}},$$

we may write for the impulse transfer function the relation

$$H_0(z, \varepsilon) \doteq \sum_{i=1}^{s} A_i' \frac{T \gamma z}{z - \mathrm{e}^{p_i T}} \mathrm{e}^{p_i T \varepsilon} \qquad \text{for} \qquad 0 < \gamma \leqq \varepsilon < 1 . \tag{3.86}$$

The same result would be obtained should we assume the impulse $d(t) = \gamma T\, \delta(t)$, $\delta(t)$ being the Dirac function.

EXAMPLE. We demonstrate the evaluation of the response of a system at an arbitrary point using the example from Subsection 3.3.1.

First, we determine the partial impulse transfer functions by relations (3.84a) and (3.84b). For $0 \leqq \varepsilon < \tau$, we obtain

$$H_A(z, \varepsilon) = 1 - \frac{z - e^{-0.2}}{z - e^{-0.5}} e^{-0.5\varepsilon} =$$

$$= \frac{z(1 - e^{-0.5\varepsilon}) - (0.606\,531 - e^{-0.5\varepsilon})}{z - 0.606\,531},$$

and for $\gamma \leqq \varepsilon < 1$ we have

$$H_B(z) = -\frac{(1 - e^{0.3})\,z}{z - e^{-0.5}} e^{-0.5\varepsilon} = \frac{0.349\,859\,z}{z - 0.606\,531} e^{-0.5\varepsilon}.$$

Since the transform of the input sequence $\{f_n\}$ is

$$F(z) = \frac{z}{z - 1},$$

we obtain, for the transform of the response,

$$F_2(z, \varepsilon) = \frac{z}{z - 1} H_A(z, \varepsilon) = \frac{z^2(1 - e^{-0.5\varepsilon}) - z(0.606\,531 - e^{-0.5\varepsilon})}{z^2 - 1.606\,531\,z + 0.606\,531}$$

for $0 \leqq \varepsilon < \gamma$, and

$$F_2(z, \varepsilon) = \frac{z}{z - 1} H_B(z, \varepsilon) = \frac{0.349\,859\,e^{-0.5\varepsilon} z}{z^2 - 1.606\,531\,z + 0.606\,531}$$

for $\gamma \leqq \varepsilon < 1$.

If we wish to obtain the formula for the sequence $\{f_{2,n}(\varepsilon)\}$, we apply Theorem 19; if we are interested in only some values of the sequence, we perform the inverse transform for the chosen ε by expansion into a series. The time behaviour of the response is indicated in Fig. 28.

3.3.3. *Evaluation of the Response at Equidistant Points*

To determine the values of the response to a sequence of rectangular impulses at equidistant points $t = mT'$ $(T' = T/i,\ i > 1$ integer, $m = 0, 1, 2, \ldots)$ we use relation (3.78)

$$F_2(z) = F(z^i)\,H(z),$$

where $H(z) = \mathscr{Z}\{h(mT')\}$, $F(z) = \mathscr{Z}\{f(nT)\}$, $F_2(z) = \mathscr{Z}\{f_2(mT')\}$.

When determining the impulse transfer function $H(z)$ the following two cases have to be distinguished:

(a) The duration of the impulse is an integer multiple of the new sampling interval T', i.e. $\tau = \gamma T' \leqq T = iT'$, $i \geqq \gamma \geqq 1$, where γ is integer.

(b) The duration of the impulse is not an integer multiple of T', i.e. $\tau = \gamma T' + \vartheta T' < T = iT'$, $0 < \vartheta < 1$, $0 \leqq \gamma < i$, γ integer.

In the first case the response to a rectangular impulse is given for $t = mT'$ by

$$h_m = h(mT') = a(mT') - a(mT' - \gamma T') = a_m - a_{m-\gamma},$$

where $a_{m-\gamma} = 0$ for $m < \gamma$. Since γ is integer, it is possible to apply the translation theorem to the $\mathscr{Z}$ transform. We obtain a single impulse transfer function

$$H(z) = \mathscr{Z}\{h_m\} = A(z) - z^{-\gamma} A(z) = \frac{z^\gamma - 1}{z^\gamma} A(z), \tag{3.87}$$

where $A(z)$ is the transform of the sequence of values of the response of the system to the unit step for $t = mT'$. If the transfer function $G_L(p)$ has simple poles none of which lies at the origin, we obtain for $A(z)$ the formula

$$A(z) = \mathscr{Z}\{a(mT)\} = \frac{A_0 z}{z - 1} + \sum_{i=1}^{s} A_i \frac{z}{z - e^{p_i T'}} \tag{3.88}$$

and for the impulse transfer function the relation

$$H(z) = A_0 \frac{z^\gamma - 1}{z^{\gamma - 1}(z - 1)} + \sum_{i=1}^{s} A_i \frac{z^\gamma - 1}{z^{\gamma - 1}(z - e^{p_i T'})}. \tag{3.89}$$

In the second case (when the duration of the impulse is not an integer multiple of the value T'), it is necessary, when applying the translation theorem, to perform the $\mathscr{Z}$ transform of the sequence obtained from the translated function. In this case, for the response of a single impulse we have the relation

$$h_m = h(mT') = a(mT') - a(mT' - \gamma T' - \vartheta T').$$

Since $a(mT' - \gamma T' - \vartheta T') = 0$ for $m = 0, 1, 2, \ldots, \gamma$, we use the modification

$$h_m = a(mT') - a[mT' - (\gamma + 1) T' + (1 - \vartheta) T'].$$

Upon application of the $\mathscr{Z}$ transform, we obtain the single impulse transfer function

$$H(z) = A(z) - z^{-(\gamma+1)} \mathscr{Z}\{a_m(1 - \vartheta)\}, \tag{3.90}$$

where

$$a_m(1 - \vartheta) = a(mT' + T' - \vartheta T').$$

If the transfer function $G_L(p)$ has simple poles none of which lies at the origin, we obtain for the impulse transfer function the relation

$$H(z) = A_0 \frac{z^{\gamma+1} - 1}{z^\gamma(z - 1)} + \sum_{i=1}^{s} A_i \frac{z^{\gamma+1} - e^{p_i T'(1-\vartheta)}}{z^\gamma(z - e^{p_i T'})}. \tag{3.91}$$

EXAMPLE. We demonstrate the evaluation of the response at equidistant points using the example of the RC two-port again. The input signal consists of rectangular impulses of unit height and width $\tau = 0.3$, with impulse period $T = 0.5$.

(a) Let us determine the values of the response at the points mT', $m = 0, 1, 2, \ldots,$ $T' = 0.1$. Since $\tau = 0.3 = 3T'$, for the impulse transfer function we obtain, with the aid of relation (3.79), for $\gamma = 3$, the relation

$$H(z) = \frac{z^3 - 1}{z^3} \frac{(1 - e^{-0.1})\, z}{(z - 1)(z - e^{-0.1})} =$$

$$= \frac{z^3 - 1}{z^3} \frac{0.095\,163}{z^2 - 1.904\,837\, z + 0.904\,837}.$$

Here we have exploited the relation

$$A(z) = \frac{(1 - e^{-0.1})\, z}{(z - 1)(z - e^{-0.1})} \qquad \text{for} \qquad T' = 0.1\,.$$

The impulse period is $T = 0.5$, thus it is also an integer multiple of the step $T' = 0.1$. Therefore, the transform of the input sequence is given by

$$F(z) = \frac{z^5}{z^5 - 1}.$$

Thus, for the transform of the output sequence we have

$$F_2(z) = \frac{z^5}{z^5 - 1} H(z) =$$

$$= \frac{0.095\,163\, z^3(z^3 - 1)}{z^7 - 1.904\,837\, z^6 + 0.904\,837\, z^5 - z^2 + 1.904\,837\, z - 0.904\,837}.$$

(b) We wish to determine the values of the response at the points $t = mT'$, $m = 0, 1, 2, \ldots,$ $T' = 0.25$. In this case the duration of the impulse is not an integer multiple of the value T'. Therefore, we use relation (3.91) for $\gamma = 1$ and $\vartheta = 0.2$:

$$H(z) = \frac{z^2 - 1}{z(z - 1)} - \frac{z^2 - \mathrm{e}^{-0.2}}{z(z - \mathrm{e}^{-0.25})} =$$

$$= \frac{0.221\,199\, z^2 - 0.181\,269\, z - 0.039\,930}{z^3 - 1.778\,801\, z^2 - 0.778\,801\, z}.$$

The period of the impulses $T = 0.5$ is twice the interval T'. Therefore, the transform of the input sequence is

$$F(z) = \frac{z^2}{z^2 - 1}.$$

Upon substitution into relation (3.78) we obtain the transform of the response

$$F_2(z) = \frac{z^2}{z^2 - 1} H(z) =$$

$$= \frac{0.221\,199\, z^3 - 0.181\,269\, z^2 - 0.039\,930\, z}{z^4 - 1.778\,801\, z^3 - 0.221\,199\, z^2 + 1.778\,801\, z - 0.778\,801}.$$

Chapter 4

Application of the $\mathscr{Z}$ Transform to the Simulation of Continuous Systems

4.1. Simulation in the Time Domain

In this section the problem of the replacement of a continuous system by a discrete system will be discussed from the point of view of the time response. The input signal for the discrete system is obtained by sampling the input signal of the continuous system. The simulation of the continuous system by the discrete system will be performed in line with three approaches which correspond to different requirements on the properties of the simulation.

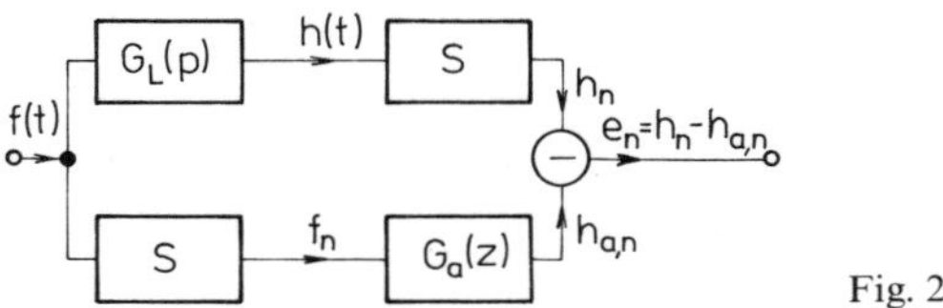

Fig. 29

Simulation in the time domain including the evaluation of the responses can be characterized by the block diagram of Fig. 29. The function $f(t)$ of the continuous variable $0 \leqq t < +\infty$ is the input signal of the continuous system with transfer function $G_L(p)$. The signal $f(t)$ is simultaneously fed into block S where sampling is executed, i.e. where the sequence $\{f_n\}$ is generated for which $f_n = f(t)|_{t=nT}$, where $T > 0$, $n = 0, 1, 2, \ldots$. Since the Dirac function is a frequent and useful input signal, we assume for the block S that sampling of the function $f(t) = \delta(t)$ yields the sequence $\{\delta_n\}$. (We shall not discuss the construction of the transfer function since this relates to the considerations of Section 3.3.) The sequence $\{f_n\}$ is the input signal of the discrete system with transfer function $G_a(z)$. The response $\{h_{a,n}\}$ should be close in a certain sense to the sampled response $\{h_n\}$ of the continuous system which is again obtained by passage of the response $h(t)$ through the block S. The difference $e_n \doteq h_n - h_{a,n}$ is used to assess the quality of the simulation of the continuous system by a discrete system. The criteria for the assessment of the sequence $\{e_n\}$ may be the following:

(a) For all $n \geqq 0$ we should have $e_n = 0$. This requirement corresponds to the interpolation method and it is satisfiable for input signals of a definite type only. The problem will be discussed in Subsection 4.1.1 for signals with Laplace transform $F_L(p) = p^{-\gamma}$ for $\gamma = 0, 1, 2$.

(b) For the absolute value of the error we should have $|e_n| \leqq E_1$, where $E_1 > 0$ for all n. This problem will be discussed in Subsections 4.1.2 and 4.1.3; it will be solved by the approximation of the convolution integral.

(c) For the sum of squares of the sequence $\{e_n\}$ we should have $\sum_{n=0}^{N-1} e_n^2 \leqq$ $\leqq E_2$, $E_2 > 0$, $0 < N \leqq +\infty$. Problems of this type, which are related for instance to the identification of systems, occur frequently in automatic regulation. One of the possible methods of solution will be given in Subsections 4.1.5 and 6.2.3.

4.1.1. *Simulation Based on the Principle of the Invariance of the Response*

The simplest approach to the problem of the simulation of the time response of a continuous system by the response of a discrete system is given by the requirement that the response of the two systems, for the given type of input signal, be identical at the points $t = nT$, $T > 0$, $n = 0, 1, 2, \ldots$. This approach was applied in Section 3.3 where the impulse transfer function of a discrete system $H(z)$ was defined as the transform of the sampled response $h(t)$ of a continuous system with transfer function $G_L(p)$ to a single impulse whose duration was $\tau < T$. If we choose $f(t) = \delta(t)$, where $\delta(t)$ is the Dirac function, as the input signal of the continuous system, then its response is equal to the impulse response and we have $h(t) =$ $= g(t) = \mathscr{L}^{-1}\{G_L(p)\}$. This delivers us from the dependence on the form of the input impulse. Then the impulse transfer function of the discrete system is given by the $\mathscr{Z}$ transform of the sampled impulse response and we have

$$H(z) = G(z) = \mathscr{Z}\{g(t)|_{t=nT}\} = \mathscr{Z}\{\mathscr{L}^{-1}\{G_L(p)\}|_{t=nT}\}\,. \tag{4.1}$$

The transfer function of the discrete system constructed in this manner is called *impulse invariant*. With reference to the further exposition we introduce for it the notation $G_0(z) = G(z)$.

Before deriving discrete transfer functions which are invariant for other signals we illustrate the procedure of the construction of $G_0(z)$ using the example of a transfer function $G_L(p)$ with simple poles. Assume that the transfer function $G_L(p)$ is of the form

$$G_L(p) = \frac{P(p)}{Q(p)},$$

where $P(p)$ is a polynomial of degree r and $Q(p)$ is a polynomial of degree s, $r = s - 1$. The transfer function is decomposed into partial fractions

$$G_L(p) = \sum_{i=1}^{s} \frac{A_i}{p - p_i}, \tag{4.2}$$

where p_i are the roots of the equation $Q(p) = 0$ and

$$A_i = \frac{P(p_i)}{\lim\limits_{p \to p_i} Q(p)/(p - p_i)} = \frac{P(p_i)}{\left.\frac{\mathrm{d}Q(p)}{\mathrm{d}p}\right|_{p=p_i}}.$$

The response of a continuous system to the unit impulse $f(t) = \delta(t)$ is given by the inverse Laplace transform (4.2) and has the form

$$g(t) = \mathscr{L}^{-1}\{G_L(p)\} = \sum_{i=1}^{s} A_i \, \mathrm{e}^{p_i t}. \tag{4.3}$$

For the response $\{g_n\}$ of the impulse invariant discrete system the relation

$$g_n = g(nT) = \sum_{i=1}^{s} A_i \, \mathrm{e}^{p_i nT}$$

must hold whence we obtain, by the $\mathscr{Z}$ transform, the desired impulse invariant transfer function

$$G_0(z) = \mathscr{Z}\{g_n\} = \sum_{i=1}^{s} A_i \frac{z}{z - \mathrm{e}^{p_i T}}. \tag{4.4}$$

The function $G_0(z)$ can again be modified to a rational function

$$G_0(z) = \frac{R(z)}{S(z)}.$$

Comparison of relations (4.2) and (4.4) shows that the poles $p = p_i$ of the function $G_L(p)$ are transformed into the poles $z = \mathrm{e}^{p_i T}$ of the function $G_0(z)$. If the poles p_i lie in the left halfplane, the poles z_i lie in the interior of the unit circle. In other words: A stable discrete system corresponds to a stable continuous system. However, no similar simple relation exists between the zeros of $G_L(p)$ and $G_0(z)$.

In this subsection we deal only with properties of systems in the time domain, whereas the discussion of the frequency properties of impulse invariant transfer functions is left to Section 4.2.

The notion of the invariance of continuous and discrete systems can be extended even to other types of signals. The input signal $f(t) = [1/(\nu - 1)!]\, t^{\nu-1}$ is advantageous and simple; its Laplace transform is $F_L(p) = 1/p^\nu$, where ν is positive integer. Thus, if the relation

$$G_L(p) = \frac{H_L(p)}{F_L(p)}$$

holds for the transfer function of a continuous system under zero initial conditions, where $H_L(p)$ is the transform of the response $h(t)$ to the input signal $f(t) = \mathcal{L}^{-1}\{F_L(p)\}$, then it is possible to define the transfer function of the discrete system by the relation

$$G(z) = \frac{H(z)}{F(z)} = \frac{\mathcal{Z}\{h(t)|_{t=nT}\}}{\mathcal{Z}\{f(t)|_{t=nT}\}} = \frac{\mathcal{Z}\{\mathcal{L}^{-1}\{H_L(p)\}|_{t=nT}\}}{\mathcal{Z}\{\mathcal{L}^{-1}\{F_L(p)\}|_{t=nT}\}}. \tag{4.5}$$

For the chosen type of input signal we have (see Subsection 2.1.8 and Section 7.3)

$$\mathcal{Z}\{f(t)|_{t=nT}\} = \mathcal{Z}\left\{\frac{T^{\nu-1}}{(\nu - 1)!}\, n^{\nu-1}\right\} = \frac{T^{\nu-1}}{(\nu - 1)!}\,\frac{N_\nu(z)}{(z - 1)^\nu}. \tag{4.6}$$

Let us substitute into relation (4.5) where we also put $H_L(p) = G_L(p)/p^\nu$. For the transfer function of a discrete system which is invariant to a signal of the $t^{\nu-1}$ type we obtain the final form

$$G_\nu(z) = \frac{(\nu - 1)!}{T^{\nu-1}}\,\frac{(z - 1)^\nu\, \mathcal{Z}\left\{\mathcal{L}^{-1}\left\{\dfrac{G_L(p)}{p^\nu}\right\}\bigg|_{t=nT}\right\}}{N_\nu(z)}. \tag{4.7}$$

The transform $G_L(p)/p^\nu$ corresponds to the ν-times repeated integration of the impulse response $g(t)$. This fact can be exploited with advantage in some physical applications where the integral of the impulse response is the characteristic function. Besides, integration of the impulse response is of advantage as well when asserting the properties of the invariant transfer function in the frequency domain (see Section 4.3).

Further, let us present invariant transfer functions for (a) $v = 1$, (b) $v = 2$.

(a) For the transfer function which is invariant to the unit step we obtain, upon substitution of $v = 1$ into (4.7),

$$G_1(z) = \frac{z-1}{z} \mathscr{Z}\left\{ \mathscr{L}^{-1}\left\{\frac{1}{p} G_L(p)\right\}\Big|_{t=nT}\right\} =$$

$$= \frac{z-1}{z} \mathscr{Z}\left\{\int_0^{nT} g(\tau)\,\mathrm{d}\tau\right\}. \tag{4.8}$$

If we assume that the transfer function with simple poles, none of which lies at the origin, is decomposed into partial fractions, then we have

$$\mathscr{L}^{-1}\left\{\frac{1}{p}\sum_{i=1}^{s}\frac{A_i}{p-p_i}\right\} = \sum_{i=1}^{s}\frac{A_i}{p_i}\left(\mathrm{e}^{p_i t} - 1\right). \tag{4.9}$$

Upon substitution into (4.7) we obtain

$$G_1(z) = \frac{z-1}{z}\mathscr{Z}\left\{\sum_{i=1}^{s}\frac{A_i}{p_i}\left(\mathrm{e}^{p_i T n} - 1\right)\right\} = \sum_{i=1}^{s}\frac{A_i}{p_i}\,\frac{\mathrm{e}^{p_i T} - 1}{z - \mathrm{e}^{p_i T}}. \tag{4.10}$$

(b) A transfer function which is invariant to linearly increasing input signals, i.e. for $v = 2$, is

$$G_2(z) = \frac{1}{T}\frac{(z-1)^2}{z}\mathscr{Z}\left\{\mathscr{L}^{-1}\left\{\frac{1}{p^2}G_L(p)\right\}\Big|_{t=nT}\right\} =$$

$$= \frac{1}{T}\frac{(z-1)^2}{z}\mathscr{Z}\left\{\int_0^{nT}\int_0^{t} g(\tau)\,\mathrm{d}\tau\,\mathrm{d}t\right\}. \tag{4.11}$$

For the transfer function with the properties mentioned in preceding relation (4.9) we obtain, upon integration of the response to the unit step and upon modification

$$G_2(z) = \frac{1}{T}\sum_{i=1}^{s}\frac{1}{p_i^2}\,\frac{B_i z + C_i}{z - \mathrm{e}^{p_i T}}, \tag{4.12}$$

where

$$B_i = A_i\left(\mathrm{e}^{p_i T} - 1 - p_i T\right), \qquad C_i = A_i\left[\mathrm{e}^{p_i T}(p_i T - 1) + 1\right].$$

Transfer functions which are invariant for signals of the $t^{\nu-1}$ type, $\nu = 3, 4, \ldots$, can be constructed similarly. However, the properties of polynomials $N_\nu(z)$ reviewed in Section 7.3 imply that for odd ν starting from $\nu = 3$ one pole of the discrete invariant transfer function always lies at the point $z = -1$, and for $\nu \geqq 4$ some of the poles lie outside the unit circle. This means that for $\nu \geqq 4$ it is necessary to interpret discrete transfer functions $G_\nu(z)$ in the sense of the two-sided $\mathscr{Z}$ transform.

EXAMPLE. To assess the significance of the choice of ν for signals of the $t^{\nu-1}$ type we determine, to a continuous system with the simple transfer function $G_L(p) = 1/(p+1)$, the behaviour of the impulse response and of the response to the signal $f(t) = t$ for invariant transfer functions $G_\nu(z)$, $\nu = 0, 1, 2$. We choose the sampling interval $T = 1$ and obtain the transfer functions

$$G_0(z) = \frac{z}{z - e^{-1}}, \qquad G_1(z) = \frac{1 - e^{-1}}{z - e^{-1}},$$

$$G_2(z) = e^{-1}\frac{z - 2 + e}{z - e^{-1}}.$$

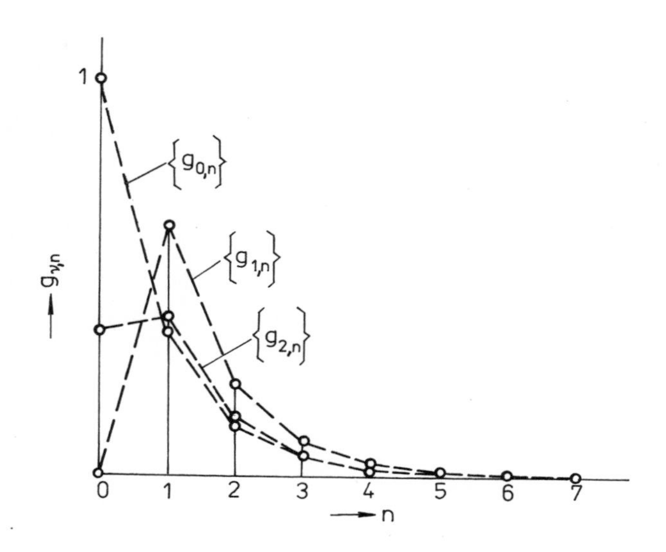

Fig. 30

For all three discrete transfer functions Fig. 30 shows the impulse responses, Fig. 31 the responses to the linearly increasing signal.

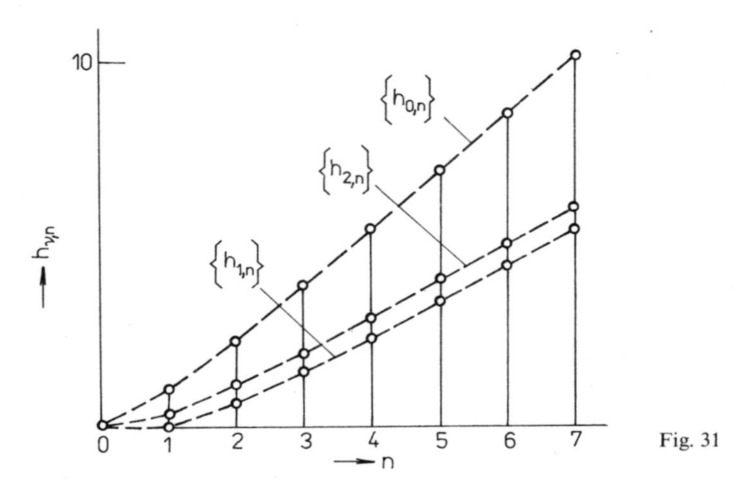

Fig. 31

4.1.2. *Approximation of the Convolution Integral*

For the derivation of the transfer function of the discrete system which simulates a continuous system it was required that the two systems be invariant for a certain type of signal at the points $t = nT$. This requirement is sometimes not satisfiable nor expedient. Therefore, we shall construct the transfer function of the discrete system under the assumption of a general signal. To this we apply the approximation of the convolution integral.

The response $h(t)$ of a continuous system with transfer function $G_L(p)$ is given by the convolution integral

$$h(t) = \int_0^t f(\tau)\, g(t - \tau)\, \mathrm{d}\tau\,, \tag{4.13}$$

where $f(t)$ is the input signal of the system and $g(t) = \mathscr{L}^{-1}\{G_L(p)\}$ is the impulse response of the system. We put $t = nT$ and approximate the integral according to the Euler–Laplace quadrature formula

$$\int_0^{nT} y(\tau)\, \mathrm{d}\tau = T\Bigg[\sum_{k=0}^{n} y_k - \frac{1}{2}(y_n + y_0) - \\ - \frac{1}{12}(\Delta y_{n-1} - \Delta y_0) - \frac{1}{24}(\Delta^2 f_{n-2} - \Delta^2 f_0) + \dots\Bigg], \tag{4.14}$$

where $y_k = y(kT)$, $\Delta^{r+1} y_k = \Delta^r y_{k+1} - \Delta^r y_k$, $\Delta^0 y_k = y_k$. If we limit ourselves to differences of order r and if the derivatives of the function $y(t)$ are continuous up to order $r + 1$, then the error of the approximation is proportional to the value T^{r+2}.

(a) *Rectangular approximation.* If we use only the first term for the approximation of the convolution integral, i.e. only the sequence of partial sums in relation (4.14), we obtain

$$h_{1,n} = T \sum_{k=0}^{n} f_k g_{n-k}, \tag{4.15}$$

where the notation $h_{1,n} = h_1(nT)$ is introduced for the approximation of the response $h(t)$ by the rectangular rule. The error of the approximation is proportional to the interval T. The sequence $\{h_{1,n}\}$ is given by convolution of the sequences $\{f_n\}$ and $\{g_n\}$. The $\mathscr{Z}$ transform is

$$H_1(z) = \mathscr{Z}\{h_{1,n}\} = T\, F(z)\, G(z),$$

where we again have $F(z) = \mathscr{Z}\{f_n\}$ and $G(z) = \mathscr{Z}\{g_n\}$. The transfer function of the corresponding discrete system is denoted by $G_1(z)$ and defined by the relation

$$G_1(z) = \frac{H_1(z)}{T\, F(z)} = G(z). \tag{4.16}$$

We have arrived again at the impulse invariant transfer function $G_0(z)$ derived in Subsection 4.1.1. Consequently, it is possible to summarize as follows: The impulse invariant transfer function of the discrete system corresponds to the rectangular approximation of the convolution integral. This approximation has error proportional to the interval T.

(b) *Trapezoidal approximation.* Now, let us use for the approximation of the convolution integral the first two terms of relation (4.14). For the sequence of approximate values of the response $\{h_{2,n}\}$ with approximation error proportional to T^2 we then have

$$h_{2,n} = T\left[\sum_{k=0}^{n} f_k g_{n-k} - \frac{1}{2}(f_n g_0 + f_0 g_n)\right], \tag{4.17}$$

where $f_0 = f(0)$ and $g_0 = g(0)$. The $\mathscr{Z}$ transform yields

$$H_2(z) = T\left\{F(z)\left[G(z) - \frac{g_0}{2}\right] - \frac{f_0}{2}\, G(z)\right\}.$$

The transform of the approximate response $H_2(z)$ is given as the sum of two terms. The left term corresponds to the solution of a certain difference equation with zero initial conditions while the right term corresponds to the solution of the same difference equation but homogeneous with initial condition f_0. Thus, the transfer function of the discrete system obtained by the trapezoidal approximation is defined as follows:

$$G_2(z) = \frac{H_2(z)}{T\,F(z)} = G(z) - \frac{g_0}{2} = G_1(z) - \frac{g_0}{2}. \tag{4.18}$$

It is obvious that for $g_0 = 0$ we have the equality $G_2(z) = G_1(z) = G(z)$. This case arises when the inequality $r < s - 1$ holds for the degrees r and s of the numerator and denominator of the function $G_L(p)$, respectively.

Now, let us construct the function $G_2(z)$ which corresponds to the function $G_L(p)$ of Subsection 4.1.1. Using result (4.4) we obtain

$$G_2(z) = \sum_{i=1}^{s} A_i \frac{z}{z - e^{p_i T}} - \frac{g_0}{2} = \sum_{i=1}^{s} \frac{A_i}{2} \frac{z + e^{p_i T}}{z - e^{p_i T}}, \tag{4.19}$$

since $g_0 = \sum_{i=1}^{s} A_i$. Then the transform $H_2(z)$ of the approximate response under the trapezoidal approximation is

$$H_2(z) = T\left[F(z)\sum_{i=1}^{s} \frac{A_i}{2} \frac{z + e^{p_i T}}{z - e^{p_i T}} - \frac{f_0}{2} \sum_{i=1}^{s} A_i \frac{z}{z - e^{p_i T}}\right]. \tag{4.20}$$

Similarly, using differences of higher orders, we determine further discrete transfer functions which approximate with different precision the transfer properties of the continuous system. However, for the majority of the applications the trapezoidal approximation with error proportional to T^2 is as a rule sufficient. This means, e.g., that using half the original step leads to the decrease of the error to $1/4$. Error estimation will be discussed in Subsection 4.1.4 below.

4.1.3. *Numerical Inverse Laplace Transform*

In the preceding two subsections, the transform $G(z)$ of the sequence $\{g_n\}$ produced by sampling the impulse response of a continuous system was obtained, as a rule, by the inverse Laplace transform of the transfer function $G_L(p)$. However, it is sometimes more advisable to apply the method of the approximate inverse Laplace transform by means of the $\mathscr{Z}$ transform: with its aid the transfer function of a continuous system $G_L(p)$ is converted to the transfer function of the discrete system $G_a(z)$ which simulates the transfer properties of the continuous system. This method consists in the numerical solution of an integral equation and exploits again the approximation of the convolution integral. For the majority of the applications the approximation given by the trapezoidal rule proves suitable. Therefore, we limit the exposition below only to this type of approximation.

Let us assume that the transfer function $G_L(p)$ is given by the ratio of two polynomials in the form

$$G_L(p) = \frac{a_1 p^{s-1} + a_2 p^{s-2} + \ldots + a_s}{p^s + b_1 p^{s-1} + \ldots + b_s}.$$

For the transform of the response we have $H_L(p) = F_L(p)\, G_L(p)$, where $F_L(p) \neq 1$. Let us substitute for $G_L(p)$ and divide the numerator as well as the denominator by p^s. We obtain

$$H_L(p) = F_L(p) \frac{A_L(p)}{1 + B_L(p)}, \tag{4.21}$$

where

$$A_L(p) = \sum_{i=1}^{s} \frac{a_i}{p^i}, \qquad B_L(p) = \sum_{i=1}^{s} \frac{b_i}{p^i}.$$

Thus we have obtained the functions $A_L(p)$ and $B_L(p)$ which are Laplace transforms. Multiply both sides of (4.21) by the expression $1 + B_L(p)$ and perform the inverse Laplace transform. The procedure yields the Volterra integral equation

$$h(t) + \int_0^t h(\tau)\, b(t - \tau)\, d\tau = \int_0^t f(\tau)\, a(t - \tau)\, d\tau, \tag{4.22}$$

where

$$a(t) = \mathscr{L}^{-1}\{A_L(p)\} = \sum_{i=1}^{s} \frac{a_i}{(i-1)!} t^{i-1}, \tag{4.23a}$$

$$b(t) = \mathscr{L}^{-1}\{B_L(p)\} = \sum_{i=1}^{s} \frac{b_i}{(i-1)!} t^{i-1}. \tag{4.23b}$$

Both integrals are approximated by the trapezoidal rule and we introduce the notation $\{h_{2,n}\}$ for the sequence of approximate values of the solution, where the index 2 symbolizes the application of the trapezoidal rule with approximation error proportional to T^2. We obtain

$$h_{2,n} + T\left[\sum_{k=0}^{n} h_{2,k}b_{n-k} - \frac{1}{2}(h_{2,0}b_n + h_{2,n}b_0)\right] =$$
$$= T\left[\sum_{k=0}^{n} f_k a_{n-k} - \frac{1}{2}(f_n a_0 + f_0 a_n)\right]. \tag{4.24}$$

By substitution into relations (4.23a, b) we easily verify that $a_0 = a(0) = a_1$, $b_0 = b(0) = b_1$.

Applying the $\mathscr{Z}$ transform and solving the equation for the transform $H_2(z)$ of the approximate response we obtain the resulting form

$$H_2(z) = F(z)\frac{T\left[A(z) - \frac{1}{2}a_1\right]}{1 + T\left[B(z) - \frac{1}{2}b_1\right]} +$$
$$+ \frac{\frac{T}{2} f_0\, A(z)}{1 + T\left[B(z) - \frac{1}{2}b_1\right]} =$$
$$= F(z)\, G_2(z) + \frac{T}{2} f_0 \frac{A(z)}{1 + T\left[B(z) - \frac{1}{2}b_1\right]}. \tag{4.25}$$

The function $G_2(z)$ is the transfer function of the discrete system which simulates the transfer properties of the continuous system. The second term in relation (4.25) corrects the response of the discrete system in the case of $f_0 \neq 0$.

Now, we direct our attention to the functions $A(z) = \mathscr{Z}\{a_n\}$ and $B(z) = \mathscr{Z}\{b_n\}$. Based on Subsection 2.1.8 we easily derive

$$A(z) = \sum_{i=1}^{s} a_i \frac{T^{i-1}}{(i-1)!} \frac{N_i(z)}{(z-1)^i},$$

$$B(z) = \sum_{i=1}^{s} b_i \frac{T^{i-1}}{(i-1)!} \frac{N_i(z)}{(z-1)^i},$$

where $N_i(z)$ are polynomials given in Section 7.3.

Upon substitution of these relations into (4.25) the transform $H_2(z)$ can be modified to a ratio of two polynomials. Its inverse transform can then be obtained algebraically according to one of the procedures presented in Subsection 2.2.2.

Till now it was assumed that the input signal of the continuous system is a general function $f(t)$ and our problem is to determine the sequence of approximate values of the response $\{h_{2,n}\}$. However, if the input signal is the impulse function $f(t) = \delta(t)$, we look for the sequence of approximate values of the impulse response $\{g_{2,n}\}$, and on the right-hand side of relation (4.22) we shall have only the functions $a(t)$ instead of the convolution integral. Again, the approximation of the integral equation and the $\mathscr{Z}$ transform are performed. The solution of the obtained equation yields the discrete transfer function

$$\tilde{G}_2(z) = \frac{A(z) + \dfrac{T}{2} a_1 B(z)}{1 + T\left[B(z) - \dfrac{1}{2} b_1\right]}. \tag{4.26}$$

The tilda in the notation of the discrete transfer function is to underline the fact that it is not identical with the function $G_2(z)$. The inverse transform of the discrete transfer function $\tilde{G}_2(z)$ by series expansion yields the sequence $\{\tilde{g}_{2,n}\}$, which approximates the impulse response $g(t)$ with error proportional to T^2, in a purely algebraic manner.

EXAMPLE. The procedure of the construction of the discrete transfer function $\tilde{G}_2(z)$ is shown on the transfer function $G_L(p)$ decomposed into partial fractions. Although this example may seem not typical due to its simplicity it helps us understand the relations between transfer functions obtained by the different simulation approaches.

Thus, let us have the transfer function $G_L(p)$ in the form

$$G_L(p) = \sum_{i=1}^{s} \frac{A_i}{p - p_i}.$$

If we apply relation (4.26) to the individual summands, we obtain

$$\begin{aligned}
\tilde{G}_2(z) &= \sum_{i=1}^{s} A_i \frac{\dfrac{z}{z-1} + \dfrac{T}{2}\left(-\dfrac{p_i z}{z-1}\right)}{1 + T\left(-\dfrac{p_i z}{z-1} + \dfrac{p_1}{2}\right)} = \\
&= \sum_{i=1}^{s} A_i \frac{z\left(1 - \dfrac{Tp_i}{2}\right)}{z\left(1 - \dfrac{Tp_i}{2}\right) - \left(1 + \dfrac{Tp_i}{2}\right)} = \\
&= \sum_{i=1}^{s} \frac{A_i z}{z - \dfrac{1 + \dfrac{Tp_i}{2}}{1 - \dfrac{Tp_i}{2}}}.
\end{aligned} \tag{4.27}$$

Comparing the obtained relation with the discrete transfer function constructed using the impulse invariance principle, we find out that they differ only in the replacement of the poles $e^{p_i T}$ by the values $(1 + Tp_i/2)/(1 - Tp_i/2)$. This fact will be encountered later in Subsection 4.2.2 when simulating a continuous system in the frequency domain.

4.1.4. *Error Estimate of Simulation in the Time Domain*

The methods described in Subsections 4.1.2 and 4.1.3 consisted in the approximation of the convolution integral with the aid of the Euler–Laplace quadrature formula (4.14). The error sequence $\{e_n\}$ which originates from this approximation is proportional to the value T^{r+2}, where T is the sampling interval and r is the order of the difference used. For the trapezoidal approximation, i.e. for $r = 0$, it can be represented for the integral of the form (4.14) by the relation

$$e_n = -\frac{T^2}{12} nT \frac{d^2}{d\tau^2} y(\tau)\bigg|_{\tau = n_0 T}, \tag{4.28}$$

where $0 \leqq n_0 \leqq n$. Although this relation enables the estimate of the upper bound of the approximation error, its application is toilsome and has almost no practical significance.

Therefore we apply a different approach which yields satisfactory results and is very simple. It consists in the comparison of the results of two approximations for different sampling intervals T from which we extrapolate a new solution whose error is proportional to at least $T^{(r+1)+2}$. Then the extrapolated solution is taken as the basis and the two former solutions are related to it. In numerical calculus this method of increasing the accuracy of the numerical solution of problems is frequently called the *Richardson extrapolation.*

Let us illustrate this approach using simulation based on the trapezoidal approximation of the convolution integral. Taking into account its simplicity, this approach yields satisfactory results in most cases. In accordance with relation (4.28) the sequence of approximation errors can be written in the form

$$e_n = T^2 \eta_n, \qquad n = 0, 1, 2, \ldots .$$

Let us assume that the sequence $\{\eta_n\}$ depends only on the integration interval nT and is not directly dependent on the sampling interval T. As a rule, this assumption is satisfied in practice with sufficient accuracy. For the approximation error for two distinct sampling intervals T_1 and T_2 at the same point $n_1 T_1 = n_2 T_2 = nT$ we then have

$$f_n - f_{1,n} = T_1^2 \eta_n, \tag{4.29a}$$
$$f_n - f_{2,n} = T_2^2 \eta_n, \tag{4.29b}$$

where $\{f_{1,n}\}$ and $\{f_{2,n}\}$ are the sequences of values of the approximate solution for $T = T_1$ and $T = T_2$, respectively. If we remove the unknown sequence $\{\eta_n\}$ from relations (4.29a) and (4.29b) and if we denote the sequence extrapolated from the two solutions by $\{f_{12,n}\}$, we obtain upon modification

$$f_{12,n} = f_{1,n} \frac{1}{1 - \left(\frac{T_1}{T_2}\right)^2} + f_{2,n} \frac{1}{1 - \left(\frac{T_2}{T_1}\right)^2}. \tag{4.30}$$

The error of the extrapolated sequence is proportional to T^3. Then the sequence of deviations $\{e_{2,n}\} = \{f_{12,n} - f_{2,n}\}$ is represented by the expression

$$e_{2,n} = f_{12,n} - f_{2,n} = \frac{\left(\frac{T_2}{T_1}\right)^2}{1 - \left(\frac{T_2}{T_1}\right)^2} (f_{2,n} - f_{1,n}). \tag{4.31}$$

The sequence $\{e_{2,n}\}$ makes it possible to estimate the accuracy of the simulation of a continuous system by a discrete system.

4.1.5. *System Identification from Its Time Response*

The problem of simulating the behaviour of a continuous system by a discrete system can be viewed as a problem of the identification of the parameters of the discrete system.

We shall apply the algebraic approach of the inverse $\mathscr{Z}$ transform to the construction of a system of linear equations whose solution yields the coefficients of the transfer function of the discrete system. This approach is related to the Prony method usually applied to the identification of the parameters of physical systems.

We start from the considerations introduced in Subsection 2.2.2 where the inverse transform was obtained of the transform $H(z)$ given by the product

$$H(z) = F(z)\,G(z).$$

Here, the function $F(z)$ is given by a series in the variable z^{-1} while $G(z)$ is given in the form of a rational function

$$G(z) = \frac{a_0 z^s + a_1 z^{s-1} + \dots + a_s}{z^s + b_1 z^{s-1} + \dots + b_s}.$$

For the sequence of values of the response $\{h_n\}$ we have, under the assumption that $f_n = 0$ for $n < 0$, the relations

$$\begin{aligned} h_0 &= a_0 f_0 , \\ h_1 &= a_0 f_1 + a_1 f_0 - b_1 h_0 , \\ &\vdots \\ h_n &= \sum_{i=0}^{s} a_i f_{n-i} - \sum_{i=1}^{s} b_i h_{n-i} \end{aligned}$$

which may be viewed as a system of equations for the unknown coefficients a_i and b_i, $i = 1, 2, \ldots, s$, if the sequences $\{f_n\}$ and $\{h_n\}$ are known. The system of equations represented in matrix notation is

$$\begin{bmatrix} f_0 & 0 & \ldots & 0 & 0 & \ldots & 0 \\ f_1 & f_0 & \ldots & 0 & -h_0 & \ldots & 0 \\ f_2 & f_1 & \ldots & 0 & -h_1 & \ldots & 0 \\ \vdots & & & & & & \\ f_s & f_{s-1} & \ldots & f_0 & -h_{s-1} & \ldots & -h_0 \\ \vdots & & & & & & \\ f_n & f_{n-1} & \ldots & f_{n-s} & -h_{n-1} & \ldots & -h_{n-s} \end{bmatrix} \begin{bmatrix} a_0 \\ a_1 \\ \vdots \\ a_s \\ b_1 \\ \vdots \\ b_s \end{bmatrix} = \begin{bmatrix} h_0 \\ h_1 \\ h_2 \\ \vdots \\ h_s \\ \vdots \\ h_n \end{bmatrix} . \quad (4.32)$$

The system of equations has a solution if the number of equations $n + 1$ is equal to the number of unknown coefficients, i.e. to $2s + 1$. If the sequences $\{f_n\}$ and $\{h_n\}$ are given by measured values and if we do not know the order of the system s, it is of advantage to choose a larger number of equations and to solve the resulting system, e.g., by the least-squares method. From the coefficients obtained by the solution of the system of equations we then construct the transfer function of the discrete system which simulates the response of the continuous system in the sense of least-squares. By repeated solutions for distinct chosen orders s, or for different numbers of equations, it is then possible to obtain a solution which satisfies the selected criterion for the error sequence $\{e_n\}$.

If this procedure is applied to the determination of the discrete transfer function from the impulse response $\{g_n\}$, we have $\{f_n\} = \{\delta_n\}$ for the input sequence and $\{h_n\} = \{g_n\}$ for the output sequence. Then the system of equations written in matrix notation is

$$\begin{bmatrix} 1 & 0 & 0 & \dots & 0 & 0 & \dots & 0 \\ 0 & 1 & 0 & \dots & 0 & -g_0 & \dots & 0 \\ 0 & 0 & 1 & \dots & 0 & -g_1 & \dots & 0 \\ \vdots & & & & & & & \\ 0 & 0 & 0 & \dots & 1 & -g_{s-1} & \cdots & -g_0 \\ \vdots & & & & & & & \\ 0 & 0 & 0 & \dots & 0 & -g_{n-1} & \cdots & -g_{n-s} \end{bmatrix} \begin{bmatrix} a_0 \\ a_1 \\ \vdots \\ a_s \\ b_1 \\ \vdots \\ b_s \end{bmatrix} = \begin{bmatrix} g_0 \\ g_1 \\ g_2 \\ \vdots \\ g_s \\ \vdots \\ g_n \end{bmatrix}. \tag{4.33}$$

System (4.33) is obtained from equation (4.32) by the substitution $f_0 = 1$, $f_n = 0$ for $n \neq 0$ and $h_n = g_n$.

The indicated procedure enables to determine the transfer function of the discrete system from the sequences $\{f_n\}$ and $\{h_n\}$ for which $f_n = h_n = 0$ for $n < 0$, thus from the transient.

However, sometimes the above assumption cannot be satisfied since both sequences are two-sided. Then we have for every term of the sequence $\{h_n\}$ the relation

$$h_n = \sum_{i=0}^{s} a_i f_{n-i} - \sum_{i=1}^{s} b_i h_{n-i} \tag{4.34}$$

for all n. Thus the resulting system of equations in matrix notation is

$$\begin{bmatrix} f_0 & f_{-1} & f_{-2} & \cdots & f_{-s} & -h_{-1} & \cdots & -h_{-s} \\ f_1 & f_0 & f_{-1} & \cdots & f_{-s+1} & -h_0 & \cdots & -h_{-s+1} \\ \vdots & & & & & & & \\ f_s & f_{s-1} & f_{s-2} & \cdots & f_0 & -h_{s-1} & \cdots & -h_0 \\ & & & & & & & \\ \vdots & & & & & & & \\ f_n & f_{n-1} & f_{n-2} & \cdots & f_{n-s} & -h_{n-1} & \cdots & -h_{n-s} \end{bmatrix} \begin{bmatrix} a_0 \\ a_1 \\ \vdots \\ a_s \\ b_1 \\ \vdots \\ b_s \end{bmatrix} = \begin{bmatrix} h_0 \\ h_1 \\ \vdots \\ h_s \\ \\ \vdots \\ h_n \end{bmatrix}. \tag{4.35}$$

The number of unknown coefficients is again $2s + 1$. The system has a unique solution if the number of equations satisfies $n + 1 = 2s + 1$ and if, simultaneously, the rank of the matrix of the system is equal to the rank of the augmented matrix of the system. If we choose a larger number of equations than the number of unknown coefficients, we solve the resulting system by the least-squares method as indicated briefly below:

If matrix equation (4.35) is written in the form

$$\boldsymbol{a} \,.\, \boldsymbol{x} = \boldsymbol{b}\,, \tag{4.36}$$

we obtain the normal symmetric system of linear equations by left multiplication of equation (4.36) by the transposed matrix ${}^{T}\boldsymbol{a}$

$${}^{T}\boldsymbol{a} \,.\, \boldsymbol{a}\boldsymbol{x} = {}^{T}\boldsymbol{a} \,.\, \boldsymbol{b}\,. \tag{4.37}$$

The resulting system of equations has the same number of equations as of unknowns, i.e. $2s + 1$. Its solution yields the coefficients of the discrete system whose response approximates the given response with least squared error.

4.2 Simulation in the Frequency Domain

In the preceding section our aim was to replace a continuous system by a discrete system such that its time response satisfies a chosen criterion. However, we frequently encounter problems in which requirements on the behaviour of the continuous system are formulated in the frequency domain by frequency characteristics. When simulating such continuous systems by discrete systems we then have to assess the quality of the simulation in the frequency domain.

First, we review the basic concepts and the notation which will be needed for the comparison of the frequency properties of continuous and discrete systems. For the sake of simplicity the characteristics of the continuous and discrete systems will be denoted by the same letters.

Transfer function

continuous systems $\qquad G_{\mathrm{L}}(p) = \mathscr{L}\{g(t)\}$,

discrete systems $\qquad G(z) = \mathscr{Z}\{g_n\}$.

Impulse response

continuous systems $g(t) = \mathscr{L}^{-1}\{G_L(p)\}$,

discrete systems $\{g_n\} = \mathscr{Z}^{-1}\{G(z)\}$.

Complex frequency response

continuous systems

$$G_L(p)|_{p=j\omega} = G_L(j\omega) = U(\omega) + j\,V(\omega),$$
$$G_L(j\omega) = A(\omega)\,e^{j\varphi(\omega)},$$

discrete systems

$$G(z)|_{z=e^{j\omega T}} = G(e^{j\omega T}) = U(\omega) + j\,V(\omega),$$
$$G(e^{j\omega T}) = A(\omega)\,e^{j\varphi(\omega)}.$$

Amplitude response

continuous systems $A(\omega) = |G_L(j\omega)| = \sqrt{[U^2(\omega) + V^2(\omega)]}$,

discrete systems $A(\omega) = |G(e^{j\omega T})| = \sqrt{[U^2(\omega) + V^2(\omega)]}$.

Phase characteristic

continuous systems $\varphi(\omega) = \arg G_L(j\omega)$,

discrete systems $\varphi(\omega) = \arg G(e^{j\omega T})$.

Attenuation characteristic

continuous systems $a(\omega) = -20 \log A(\omega)$,

discrete systems $a(\omega) = -20 \log A(\omega)$.

Characteristic of group delay

continuous systems

$$\tau(\omega) = -\frac{d\varphi(\omega)}{d\omega} = -\mathrm{Re}\frac{d \ln G_L(p)}{dp}\bigg|_{p=j\omega},$$

discrete systems

$$\tau(\omega) = -\frac{d\varphi(\omega)}{d\omega} = -T\,\mathrm{Re}\left[z\frac{d \ln G(z)}{dz}\right]\bigg|_{z=e^{j\omega T}}$$

Continuous as well as discrete systems, which are to satisfy requirements in the frequency domain from the standpoint of the selection or suppression of some frequency bands, are divided into lowpass filters, bandpass filters, highpass filters, band rejection filters, etc. As a rule, the requirements on the transfer properties of the system are formulated by means of systems of inequalities. The inequalities can be presented graphically in so-called *tolerance schemes.* For illustration, in Fig. 32 a tolerance scheme for the behaviour of the module of the transfer function of a system manifesting the properties of a lowpass filter is given. Simultaneously, it clarifies the meaning of the terminology used for the basic characteristic frequency bands.

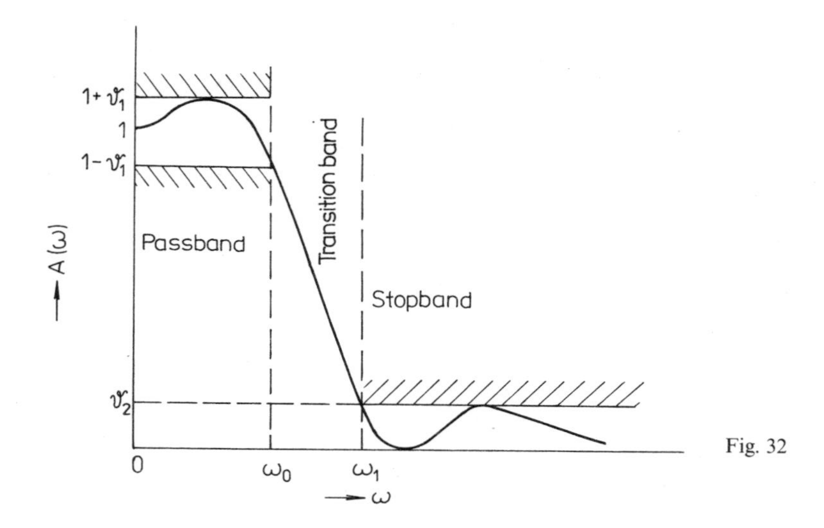

Fig. 32

The passband is defined by the requirement $1 - \vartheta_1 \leqq A(\omega) \leqq 1 + \vartheta_1$ for $|\omega| \leqq \omega_0$. On the contrary, for the stopband we require the fulfilment of the inequality $A(\omega) \leqq \vartheta_2$ for $\omega_1 \leqq \omega \leqq \omega_{\max}$. For continuous systems we usually have $\omega_{\max} = +\infty$, for discrete systems we have $\omega_{\max} = \pi/T$ because of the periodicity of the frequency response. The transition band is limited by the inequality $\omega_0 \leqq \omega \leqq \omega_1$ while the amplitude response passes from the passband to the stopband in this band.

Simulating a continuous system by a discrete system we require, as a rule, that the observed characteristic of the discrete system, e.g. the amplitude response, does not deflect from the tolerance scheme required for the behaviour of the amplitude response of the continuous system. From this standpoint the problem of the simulation of a continuous system may be viewed as an approximation problem which is solvable by specifical mathematical tools. However, in our considerations we shall start from the assumption that the transfer function of the continuous system $G_L(p)$ is given and that our aim is to design such a transfer function $G(z)$ whose frequency properties satisfy the tolerance field for the frequency response of the continuous system.

4.2.1. *Frequency Properties of Signal Invariant Transfer Functions*

To understand the mutual relation between the transfer function $G_L(p)$ of a continuous system and the transfer function $G(z)$ of a signal invariant discrete system, and also between the corresponding frequency characteristics, the following Theorem 24 is important. It deals with the relationship between the Laplace transform of the function $f(t)$ and the $\mathscr{Z}$ transform of the sequence $\{f_n\}$ obtained by sampling the function $f(t)$.

THEOREM 24. *Let $f(t)$ be a function differentiable for $t > 0$ for which we have $f(t) = 0$ for $t < 0$. Let $\lim_{t \to 0+} f(t) = f_0$. If the integral*

$$\int_0^{+\infty} |f'(t)| \, e^{-p_0 t} \, dt$$

converges for the real number $p_0 > 0$ and if

$$F_L(p) = \int_0^{+\infty} f(t) \, e^{-pt} \, dt \,, \tag{4.38}$$

then for arbitrary $T > 0$ we have the equality

$$\sum_{n=0}^{+\infty} f(nT) \, e^{-npT} = \frac{f_0}{2} + \frac{1}{T} \sum_{k=-\infty}^{+\infty} F_L\left(p + j\frac{2\pi}{T}k\right) \tag{4.39}$$

under the assumption that the series $\sum_{k=-\infty}^{+\infty} F_L\left(p + j\frac{2\pi}{T}k\right)$ converges.

If we put $e^{pT} = z$ on the left-hand side of relation (4.39), we obtain the definitorical relation of the $\mathscr{Z}$ transform which converges for $|z| > |z_0|$. On the other hand, relation (4.38) defines the Laplace transform of the function $f(t)$ which exists for $\operatorname{Re} p > p_0$. Consequently, Theorem 24 can briefly (without repeating the existence assumptions) be written as a relation between the $\mathscr{Z}$ transform and the Laplace transform:

$$F(z)\Big|_{z = e^{pT}} = \frac{f_0}{2} + \frac{1}{T} \sum_{k=-\infty}^{\to\infty} F_L\left(p + j\frac{2\pi}{T}k\right), \tag{4.40}$$

where

$$F(z) = \sum_{n=0}^{+\infty} f_n z^{-n} \qquad \text{and} \qquad F_L(p) = \mathscr{L}\{f(t)\}.$$

Let us apply Theorem 24 to the elucidation of the relationship between continuous and discrete systems. If $g(t)$ is the impulse response of the continuous system with transfer function $G_L(p)$, then, according to Subsection 4.1.1, $G(z) = \mathscr{Z}\{g_n\}$, $g_n = g(t)|_{t=nT}$, is the transfer function of the impulse invariant discrete system $G_0(z)$. By Theorem 24, the relation

$$G_0(z)|_{z=e^{pT}} = \frac{g_0}{2} + \frac{1}{T} \sum_{k=-\infty}^{+\infty} G_L\left(p + j\frac{2\pi}{T}k\right) \tag{4.41}$$

is valid which offers a direct approach to the construction of the transfer function of the impulse invariant discrete system from the known transfer function $G_L(p)$ of the continuous system with no necessity of calculating the impulse response. If we put $p = j\omega$ in relation (4.41), it is possible to construct the complex frequency response of the discrete system by the infinite sum of complex frequency characteristics of the continuous system mutually translated by multiples of the radian sampling frequency $\omega_s = 2\pi/T$. This phenomenon is called *frequency folding*, or more frequently *aliasing*.

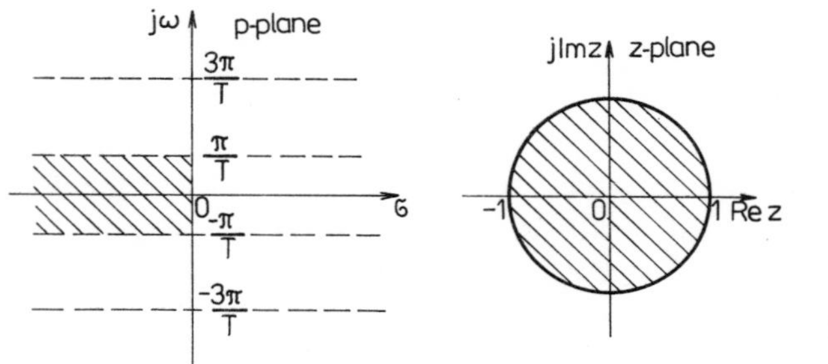

Fig. 33

From Theorem 24 it follows that the variables p and z are bound by the relation $z = e^{pT}$. This means that strips of the plane p of width $2\pi T = \omega_s$, parallel with the real axis σ, are conformly mapped into the entire z-plane (see Fig. 33). The left part of the strips for $\sigma < 0$ is mapped into the interior of the unit circle, while the part of the strips for $\sigma > 0$

is mapped onto the exterior of the unit circle. The part of the imaginary axis given by $-\pi/T < \omega < \pi/T$ is mapped into the unit circle. Thus, it is obvious that the entire p-plane is mapped into an infinite number of foils of the Riemann surface of the function $z = e^{pT}$. The function (4.41) expresses the fact that the individual p-plane strips map into overlapping foils in the z-plane, and that for systems constructed using the principle of impulse invariance a simple and unique mapping of the p-plane into the z-plane cannot be achieved.

If we substitute $p = j\omega$ in relation (4.41), the relation between frequency characteristics of the impulse invariant discrete and continuous systems is obtained:

$$G_0(e^{j\omega T}) = \frac{g_0}{2} + \frac{1}{T}\sum_{k=-\infty}^{+\infty} G_L\left(j\omega + j\frac{2\pi}{T}k\right). \tag{4.42}$$

If we have

$$G_L(j\omega) = 0 \qquad \text{for} \qquad |\omega| \geqq \frac{\pi}{T} = \frac{\omega_s}{2} \tag{4.43}$$

for the complex frequency response of the continuous system, then overlapping of the individual mutually translated components in relation (4.41) does obviously not occur and we have

$$G_0(e^{j\omega T}) = \frac{g_0}{2} + \frac{1}{T}G_L(j\omega) \tag{4.44}$$

for $|\omega| \leqq \pi/T$. The outlined considerations are in agreement with the Shannon–Kotelnikov sampling theorem according to which the function $g(t)$ is fully determined by the discrete values $g_n = g(nT)$, $n = 0, 1, 2, ..$, if we have $G_L(j\omega) = 0$ for $\omega > \omega_{max}$ and if we choose the sampling interval $T \leqq \pi/\omega_{max}$ (represented by the radian frequency $\omega_s = 2\pi/T \geqq 2\omega_{max}$).

Unfortunately, the majority of systems occurring in practice do not satisfy this assumption, but by the suitable choice of the sampling interval T it is possible to achieve at least approximate fulfilment of condition (4.43). In Fig. 34 we find the amplitude and phase response

$$G_L(p) = \frac{1}{p+1} \qquad \text{and} \qquad G_0(z) = \frac{z}{z - e^{-T}} = G(z)$$

of the impulse invariant continuous and discrete systems, respectively, for $T_1 = 1$ and $T_2 = 0.5$. The distortion of the behaviour of the frequency response of the discrete system is evident from the figures. However, this distortion decreases with the increase of the sampling frequency $f_s = 1/T$.

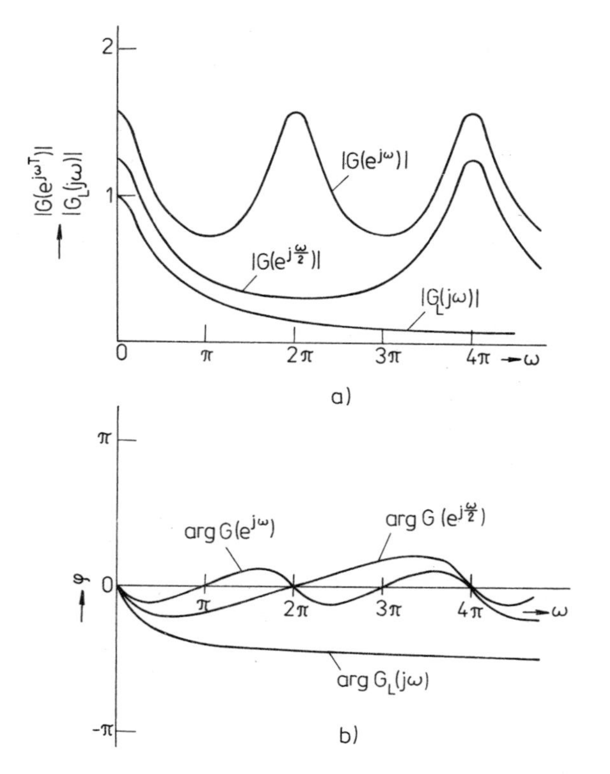

Fig. 34

The comparison of the frequency properties of the continuous system and of the discrete impulse invariant system leads thus to the important conclusion that the agreement of the impulse responses does not guarantee the agreement of the frequency characteristics.

Now, we apply Theorem 24 to the investigation of the frequency properties of the transfer functions of discrete systems which are invariant to the unit step, and to a linearly increasing function. The response of a continuous system with transfer function $G_L(p)$ to the unit step is given by the integral of its impulse response; similarly, the response to a linearly increasing signal is given by the integral of the response to the unit step. Consequently, for the responses we have

$$h_1(t) = \int_0^t g(\tau)\, \mathrm{d}\tau\,, \qquad h_1(0) = 0\,,$$

$$h_2(t) = \int_0^t \int_0^{t'} g(\tau)\, \mathrm{d}\tau\, \mathrm{d}t'\,, \qquad h_2(0) = 0\,;$$

the corresponding Laplace transforms are

$$H_{\mathrm{L}1}(p) = \frac{1}{p}\, G_{\mathrm{L}}(p)\,, \qquad H_{\mathrm{L}2}(p) = \frac{1}{p^2}\, G_{\mathrm{L}}(p)\,.$$

The convergence conditions of Theorem 24 are obviously satisfied for the functions $h_1(t)$ and $h_2(t)$ under the assumption that they are satisfied for the impulse response $g(t)$. The relation between the $\mathscr{Z}$ transform and the Laplace transform for the response to the unit step is

$$H_1(z)\big|_{z=\mathrm{e}^{pT}} = \frac{1}{T} \sum_{k=-\infty}^{+\infty} \frac{G_{\mathrm{L}}\left(p + \mathrm{j}\dfrac{2\pi}{T}k\right)}{p + \mathrm{j}\dfrac{2\pi}{T}k}\,, \tag{4.45}$$

where

$$H_1(z) = \mathscr{Z}\{h_{1,n}\}\,, \qquad h_{1,n} = h_1(t)\big|_{t=nT}\,.$$

The relation for the response to a linearly increasing input signal is

$$H_2(z)\big|_{z=\mathrm{e}^{pT}} = \frac{1}{T} \sum_{k=-\infty}^{+\infty} \frac{G_{\mathrm{L}}\left(p + \mathrm{j}\dfrac{2\pi}{T}k\right)}{\left(p + \mathrm{j}\dfrac{2\pi}{T}k\right)^2}\,, \tag{4.46}$$

where

$$H_2(z) = \mathscr{Z}\{h_{2,n}\}\,, \qquad h_{2,n} = h_2(t)\big|_{t=nT}\,.$$

The substitution $p = \mathrm{j}\omega$ yields the frequency behaviour of the corresponding responses. Now, we investigate the consequence of the integration of the impulse response upon the frequency behaviour of the functions $G_{\mathrm{L}}(p)/p$ and $G_{\mathrm{L}}(p)/p^2$. The frequency response $G_{\mathrm{L}}(\mathrm{j}\omega)$ is nonzero in the entire interval $0 \leqq |\omega| < +\infty$. However, there exists a number $0 \leqq |\omega_0|$ such that for all $|\omega| > |\omega_0|$ the amplitude response $A(\omega) = |G_L(j\omega)|$ tends monotonously to zero for increasing ω, namely with the power $\omega^{-\varkappa}$, $\varkappa > 0$. Then the frequency behaviour of $A(\omega)/\omega$ and $A(\omega)/\omega^2$ also tends to zero monotonously for increasing $|\omega|$, namely in the region not less than $|\omega| > |\omega_0|$, with powers $\omega^{-(\varkappa+1)}$ and $\omega^{-(\varkappa+2)}$, respectively – thus more rapidly than the amplitude response $A(\omega)$. The mutual overlapping of the complex frequency responses $H_{L1}(p)$ and $H_{L2}(p)$ translated by multiples of $2\pi/T$, will be smaller than the overlapping given by relation (4.42). As a consequence the deviations of the complex frequency responses $H_1(e^{j\omega T})$ and $H_2(e^{j\omega T})$ from $H_L(j\omega)$ are smaller. Fig. 35 shows the frequency behaviour of $|H_{L1}(j\omega)|$ and $|H_1(e^{j\omega})|$, while Fig. 36 brings the behaviour of $|H_{L2}(j\omega)|$ and $|H_2(e^{j\omega})|$, for the transfer function $G_L(p) = 1/(p+1)$ for $T = 1$.

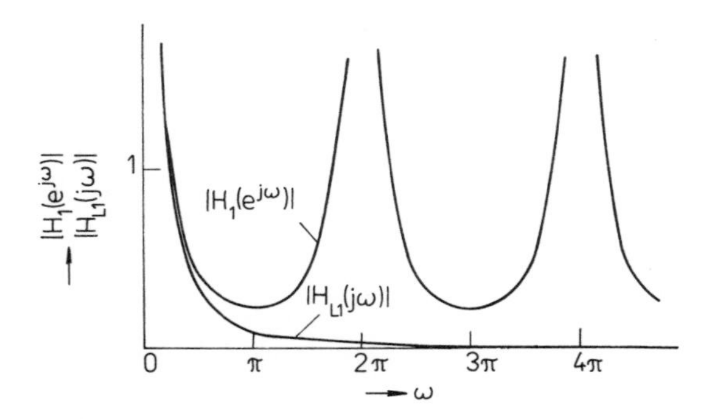

Fig. 35

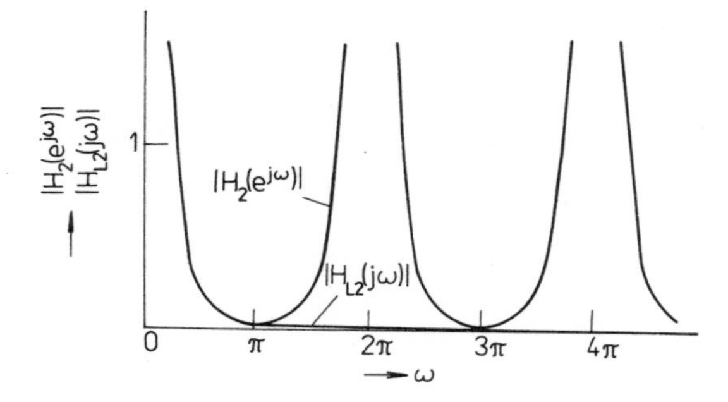

Fig. 36

In Subsection 4.1.1 the discrete transfer function $G_\nu(z)$ invariant to signals of the $t^{\nu-1}$ type was defined by the relation

$$G_\nu(z) = \frac{H_\nu(z)}{F_\nu(z)} .$$

If we substitute for $z = e^{j\omega T}$, we obtain the corresponding complex frequency response

$$G_\nu(e^{j\omega T}) = \frac{(\nu-1)!}{T^{\nu-1}} \frac{(e^{j\omega T}-1)^\nu H_\nu(e^{j\omega T})}{N_\nu(e^{j\omega T})} . \tag{4.47}$$

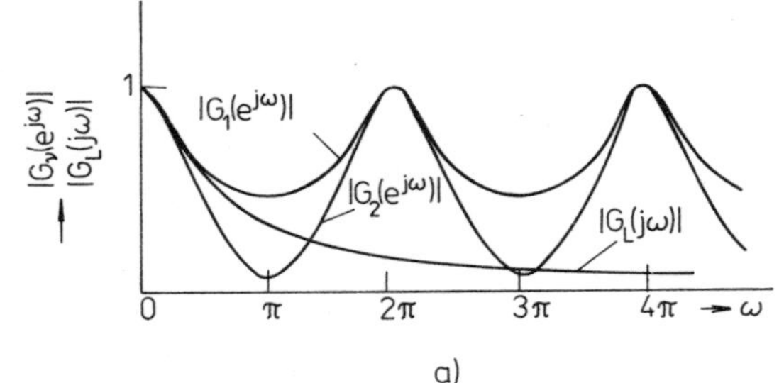

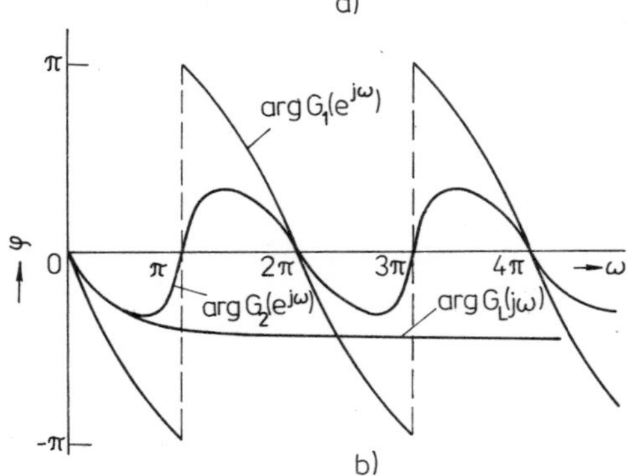

Fig. 37

The amplitude and phase responses are presented in Fig. 37a, b for the selected transfer function $G_L(p)$ for $\nu = 1, 2$. We see that integration of the impulse responses shows itself favourably on the behaviour of the amplitude as well as phase responses, and that thus discrete transfer functions invariant to signals of the $t^{\nu-1}$ type simulate better, with increasing ν, the frequency properties of the continuous system. Discrete transfer functions invariant to signals of the $t^{\nu-1}$ type can be used in cases where simple impulse invariance yields an excessively large error in the frequency domain, and where we cannot improve the properties by decreasing the interval T.

To the method of invariance to signals of the type $f(t) = t^{\nu}$ we shall return in Subsection 4.3.1 when constructing transfer functions of discrete integrators.

4.2.2. *Bilinear Transformation*

When simulating the behaviour of a continuous system with transfer function $G_L(p)$ by a discrete system with transfer function $G(z)$ in the frequency domain our aim is to determine the functions $G(z)$ so that, e.g., the behaviour of $|G(e^{j\omega T})|$ satisfies the tolerance scheme for the amplitude response $|G_L(j\omega)|$. From the considerations carried out till now it is evident that in consequence of the periodicity of the frequency characteristics of the discrete system it is not possible to require the fulfilment of the tolerance scheme for all ω but only for the values $0 \leqq \omega < \pi/T$, where T is a suitably selected sampling interval.

Theorem 24 which deals with the relationship between the Laplace transform and the $\mathscr{Z}$ transform implies that the complex variable z can be represented with the aid of the complex variable $p = \sigma + j\omega$ by the relation $z = e^{pT}$. Consequently, it should be possible to transform the transfer functions $G_L(p)$ onto the transfer functions $G(z)$ with the aid of the substitution $p = T^{-1} \ln z$. But we would obtain transcendental functions without practical significance since all the convenient algorithms for the analysis of discrete systems are based on the representation of $\mathscr{Z}$ transforms in the form of rational functions in the variable z. But it is possible to expand the function $\ln z$ into a continued fraction whose approximants are rational functions in the variable z:

$$\ln z = \cfrac{2(z-1)}{z + 1 - \cfrac{(z-1)^2}{3(z+1) - \cfrac{4(z-1)^2}{5(z+1) - \cfrac{\vdots}{\ddots \; \cfrac{(k-1)^2}{2k-1}\,\cfrac{(z-1)^2}{(z+1) - \cfrac{}{\vdots}}}}}} \tag{4.48}$$

Taking the first term of the expansion into the continued fraction yields an approximate representation for the variable $p = \sigma + j\omega$ for which we introduce the following notation:

$$w = u + jv = \frac{2}{T}\frac{z-1}{z+1}. \tag{4.49}$$

The imaginary part of the complex variable w, i.e. $\operatorname{Im} w = v$, will be called the *transformed radian frequency* in the sequel. In the literature which discusses discrete systems, relation (4.49) is called the *bilinear transformation.*

The function (4.49) belongs to the only type of functions which map uniquely the entire complex z-plane on the entire complex w-plane. The properties of the bilinear transformation will be intuitively demonstrated on the solution of (4.49) for the variable z. Let us substitute for the individual complex variables as follows:

$$z = x + jy = \frac{\frac{2}{T} + w}{\frac{2}{T} - w} = \frac{\frac{2}{T} + u + jv}{\frac{2}{T} - u - jv}. \tag{4.50}$$

For $w = jv$ we easily verify that $|z| = 1$ holds. Thus, the imaginary axis of the w-plane maps onto the unit circle, with the point $w = 0$ mapped into the point $z = 1$ and $v = +\infty$ into the point $z = -1$. The points

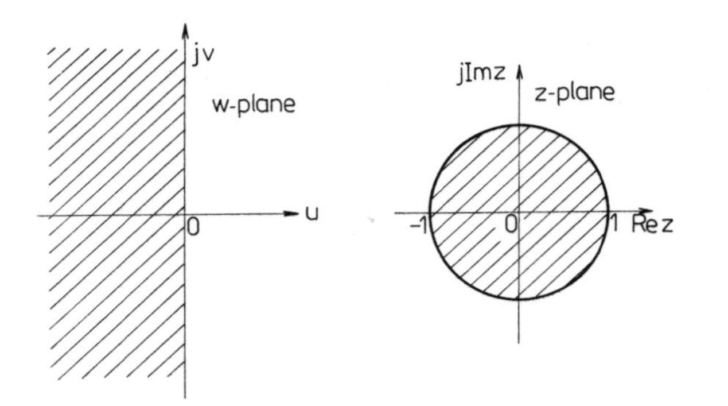

Fig. 38

of the w-plane for which we have $u < 0$ map into the unit circle $|z| < 1$, while the points in the right halfplane w, i.e. for $u > 0$, are mapped into the domain $|z| > 1$, i.e. into the exterior of the unit circle.

Thus, the bilinear transformation converts stable continuous systems to stable discrete systems and the entire imaginary axis w onto the unit circle in the z-plane. This is just the point of the basic and substantial difference between the bilinear transformation and the method of impulse invariance, for which only the segment $-\pi/T < \omega < \pi/T$ of the imaginary axis is mapped onto the unit circle. The mapping of the w-plane into the z-plane established by the bilinear transformation is shown in Fig. 38.

We gain insight into the transformation of the radian frequency ω onto the transformed radian frequency v by substituting $z = e^{j\omega T}$ into relation (4.49):

$$w = \frac{2}{T} \frac{e^{j\omega T} - 1}{e^{j\omega T} + 1} = j \frac{2}{T} \tan \frac{\omega T}{2} = u + jv . \tag{4.51}$$

The comparison of the real and imaginary parts yields

$$u = 0 , \qquad v = \frac{2}{T} \tan \frac{\omega T}{2} .$$

The dependence of the transformed radian frequency v on the radian frequency ω is shown in Fig. 39 for $T = 1$. From relation (4.51) and Fig. 39 it is obvious that the positive part of the imaginary $j\omega$ axis in the p-plane is mapped onto the positive imaginary jv axis in the w-plane and onto the upper half of the unit circle in the z-plane. On the other hand, the negative part of the $j\omega$ axis is mapped onto the negative part of the imaginary jv axis and onto the lower half of the unit circle in the z-plane. Thus, equation (4.51) can be viewed as the transformation relation between the actual radian frequency ω and the transformed radian frequency v.

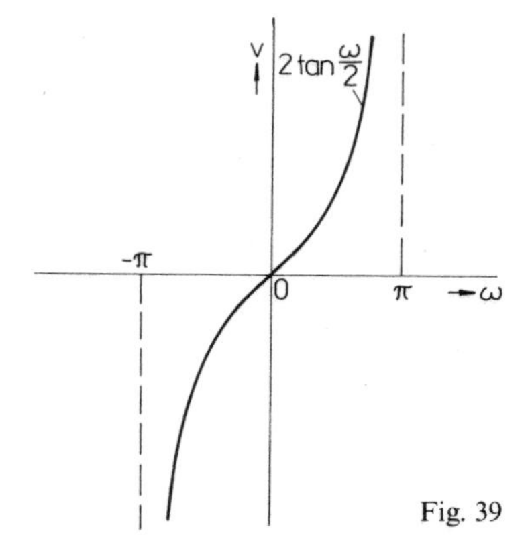

Fig. 39

A certain disadvantage of the bilinear transformation consists in the fact which follows from relation (4.51), namely that the variable v is a nonlinear function of the radian frequency ω. This implies that the bilinear transformation distorts the frequency characteristics and only for small values of ω can the function (4.51) be considered to be approximately linear. However, possibilities exist of compensating for this influence of the distortion.

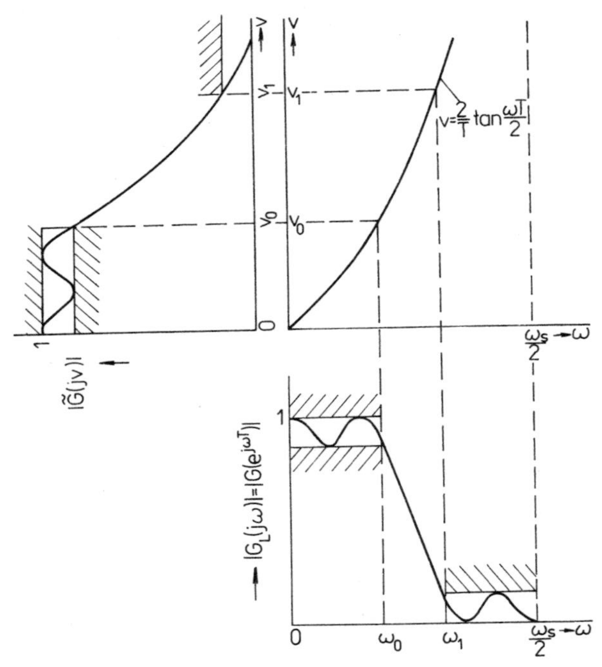

Fig. 40

The application of the bilinear transformation is simplest to the construction of a discrete transfer function $G(z)$ such that for its amplitude response the requirements are constant for certain frequency bands. For this case Fig. 40 shows the procedure of the compensation of the distortion of the frequency dependence of the amplitude response by the modification of the tolerance scheme for a discrete system of the lowpass filter type. We proceed so that we transform the characteristic frequencies of the required frequency response of the continuous system ω_0, ω_1 by relation (4.51) to the frequencies v_0, v_1. These points together with the required values of the amplitude response define the behaviour of the transformed requirements which are satisfied with the aid of a suitable approximation or frequency transformation. The result is the rational function $\tilde{G}(w)$ whose transformation to the discrete transfer function $G(z)$ will be described upon deriving the functions $\tilde{G}(w)$ for the remaining two cases.

Somewhat more complicated is the application of the bilinear transformation to the construction of a discrete transfer function with linearly increasing amplitude response where the compensation of the distortion requires still some approximation in the w-plane. For instance, if we require the behaviour of the amplitude response $A(\omega) = K\omega$, then the application of the inverse transform

$$\omega = \frac{2}{T} \arctan \frac{Tv}{2} \tag{4.52}$$

yields

$$A(\omega) = K\omega = K\frac{2}{T} \arctan \frac{Tv}{2} = \tilde{A}(v),$$

whence we obtain

$$\tilde{A}(v) = K\frac{2}{T} \arctan \frac{Tv}{2}. \tag{4.53}$$

In the transformed w-plane we then have to construct a function $\hat{G}(w)$ such that

$$|\tilde{G}(w)| = \tilde{A}(v) = K\frac{2}{T} \arctan \frac{Tv}{2}.$$

A similar procedure has to be followed in case that we require constant group delay. For group delay we have

$$\tau(\omega) = -\frac{\mathrm{d}\varphi(\omega)}{\mathrm{d}\omega}.$$

Substituting for ω by (4.52) and expressing the differential $\mathrm{d}\omega$ we obtain

$$\tau(\omega) = -\frac{\mathrm{d}\tilde{\varphi}(v)}{\mathrm{d}v}\left[1 + \left(\frac{vT}{2}\right)^2\right].$$

Since

$$-\frac{\mathrm{d}\tilde{\varphi}(v)}{\mathrm{d}v} = \tilde{\tau}(v)$$

holds for group delay in the transformed system with the transfer function $\tilde{G}(w)$, we have

$$\tilde{\tau}(v) = -\frac{\mathrm{d}\tilde{\varphi}(v)}{\mathrm{d}v} = \frac{\tau\left(\dfrac{2}{T}\arctan\dfrac{Tv}{2}\right)}{1+\left(\dfrac{vT}{2}\right)^2}. \tag{4.54}$$

Therefore, for constant group delay of the continuous as well as the discrete systems we have to construct a transfer function $\tilde{G}(w)$ such that its group delay would be

$$\tilde{\tau}(v) = \frac{1}{1+\left(\dfrac{vT}{2}\right)^2}. \tag{4.55}$$

Now, we execute the transform of the function $\tilde{G}(w)$ to the discrete transfer function $G(z)$. Let us assume a function $\tilde{G}(w)$ in the form

$$\tilde{G}(w) = K\frac{\prod_{\mu=1}^{r}(w-w_\mu)}{\prod_{\nu=1}^{s}(w-w_\nu)}.$$

If we substitute for $w = (2/T)(z-1)/(z+1)$, we obtain upon modification

$$G(z) = K\left(\frac{2}{T}\right)^{r-s}\frac{\prod_{\mu=1}^{r}\left(1-\dfrac{Tw_\mu}{2}\right)(z+1)^{s-r}\prod_{\mu=1}^{r}(z-z_\mu)}{\prod_{\nu=1}^{s}\left(1-\dfrac{Tw_\nu}{2}\right)\qquad\prod_{\nu=1}^{s}(z-z_\nu)}, \tag{4.56}$$

where

$$z_\mu = \frac{1+\dfrac{Tw_\mu}{2}}{1-\dfrac{Tw_\mu}{2}} \quad \text{and} \quad z_\nu = \frac{1+\dfrac{Tw_\nu}{2}}{1-\dfrac{Tw_\nu}{2}}.$$

EXAMPLE. To illustrate the procedure of the construction of the transfer function for the discrete system which simulates the amplitude response of a continuous system, we use the transfer function

$$G_L(p) = \frac{1}{p + 1}.$$

We choose the sampling interval $T = 1$. The corresponding amplitude response is of the type of a lowpass filter and we have

$$A(\omega) = |G_L(j\omega)| = \frac{1}{\sqrt{(1 + \omega^2)}}.$$

Let us require that the equality $|G_L(j\omega)| = |G(e^{j\omega})|$ holds at the points $\omega = 0$ and $\omega = \omega_0 = 1$. Thus, the requirements on the function $\tilde{G}(w)$ are the following:

$$|\tilde{G}(0)| = A(0) = 1 \quad \text{and} \quad |\tilde{G}(jv_0)| = A(\omega_0) = 0.707\,1\,.$$

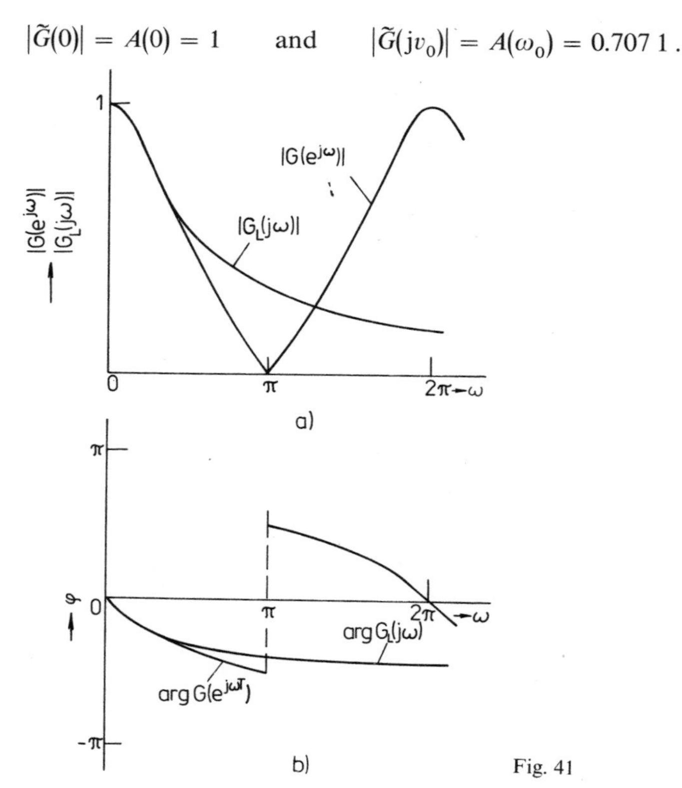

Fig. 41

With the aid of relation (4.52) we determine $v_0 = 2 \tan 0.5 = 1.092\,6$. The function $\tilde{G}(w)$ will be of the form

$$\tilde{G}(w) = \frac{1.092\,6}{w + 1.092\,6}.$$

By substitution for w and modification, or with the aid of relation (4.56), we obtain the resulting discrete transfer function

$$G(z) = 0.353\,3\,\frac{z + 1}{z - 0.293\,41}.$$

In Fig. 41a, b the amplitude and phase responses of the continuous and discrete systems are plotted, in Fig. 42 we have the corresponding impulse response. Obviously, the figures manifest good agreement of both the amplitude as well as the phase response in the required band, while the properties of the impulse response differ considerably.

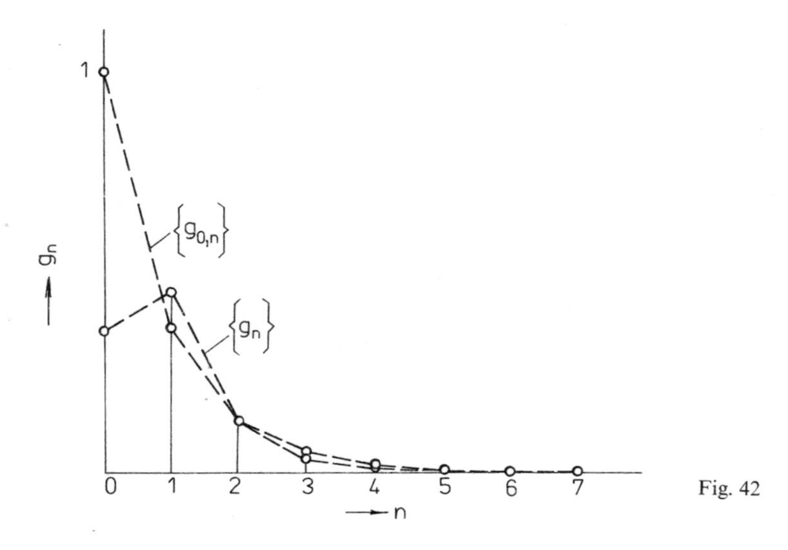

Fig. 42

4.3. Application of the $\mathscr{Z}$ Transform in Numerical Calculus

The solution of a number of problems by methods of numerical analysis requires work with sequences of values of functions, processed according to different algorithms. Possibly the best known and most often occurring problem of numerical calculus is the approximate computation of the integral. As examples of further methods which deal with sequences it is possible to mention numerical differentiation, the solution of differential equations, the solution of integral equations, etc. Some of these approaches were encountered already in Section 4.1 when simulating general continuous systems by discrete systems.

The $\mathscr{Z}$ transform is a functional transform of sequences which makes it possible to map objects from the time domain to their transforms in the complex variable z. There, certain characteristic functions correspond to operations with sequences. From the previous sections we are familiar with the concept of frequency characteristics of discrete systems. They were determined from the $\mathscr{Z}$ transforms simply by the substitution $z = e^{j\omega T}$. The variable ω is the radian frequency while $T > 0$ is the sampling interval suitably chosen for the generation of sequences of function values for continuous variables.

Starting from the fact that some problems of numerical calculus are constituted by operations on sequences it is possible to apply the $\mathscr{Z}$ transform to these operations; consequently, it is possible to gain new insight of their properties. Evaluating transforms of operations, e.g., in the frequency domain, we obtain a qualitatively different opinion on the possibilities of their application to the solution of various problems. This assumption is based upon experience with the application of the Laplace or the Fourier transforms to the solution of linear differential equations. These transforms do not lead merely to the algebraization of the solution of differential equations but they make possible also the investigation of a number of properties of the systems which they describe, e.g., by the method of zeros and poles or by means of frequency characteristics. Thus, these transforms are independent and effective tools of system theory.

In this section we apply the $\mathscr{Z}$ transform first to the representation of some familiar quadrature formulae. Further, we indicate the construction of some new transfer functions of discrete integrators which consists in the methods of simulation of continuous systems in the time and frequency domains. In the next subsection we apply the $\mathscr{Z}$ transform to the construction of transfer functions which correspond to algorithms of numerical differentiation. In all cases the properties of the transforms of operations on sequences will be considered in the frequency domain.

4.3.1. *Numerical Integration*

Assume that a function $f(t)$ is given and that we wish to determine its integral

$$h(t) = \int_0^t f(\tau)\,\mathrm{d}\tau$$

from the sequence of values of the function $f(t)$ for $t = nT$, $n = 0, 1, 2, \ldots$, $T > 0$. Obviously, we will not obtain the function $h(t)$ for all t, but only for $t = nT$. Thus, it is possible to write

$$h_n = h(nT) = \int_0^{nT} f(\tau)\,\mathrm{d}\tau\,. \tag{4.57}$$

To simplify the algorithmization of the procedure of numerical integration we divide, as a rule, the entire interval $[0, t]$ in which we integrate into equal subintervals. In these subintervals we then use simpler quadrature formulae. Relation (4.57) can then be divided into two parts:

$$h_n = \int_0^{(n-m)T} f(\tau)\,\mathrm{d}\tau + \int_{(n-m)T}^{nT} f(\tau)\,\mathrm{d}\tau\,.$$

This makes it possible to express the difference of the sequence $\{h_n\}$ over m terms by the relation

$$h_n - h_{n-m} = \int_{(n-m)T}^{nT} f(\tau)\,\mathrm{d}\tau = i_{m,n}\,. \tag{4.58}$$

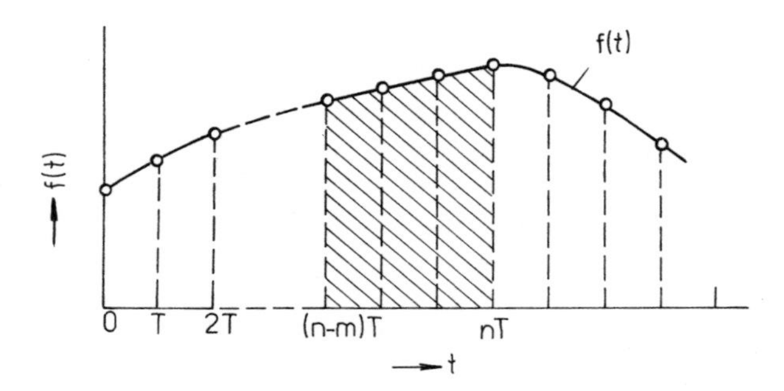

Fig. 43

The procedure is illustrated by Fig. 43, where the hatched region represents the value of the partial integral $i_{m,n}$. Applying the $\mathscr{Z}$ transform to relation (4.58) we obtain, upon modification, the following transform of the sequence of values of the integral $\{h_n\}$:

$$H(z) = \frac{z^m}{z^m - 1} I_m(z), \tag{4.59}$$

where $H(z) = \mathscr{Z}\{h_n\}$ and $I_m(z) = \mathscr{Z}\{i_{m,n}\}$. Now, we express the sequence of partial integrals $\{i_{m,n}\}$ by means of the sequence $\{f_n\}$ for which we have $f_n = f(nT)$. The value m will be called the order of the quadrature formula. In the simplest case we have $m = 1$, and using the rectangular rule we obtain

$$i_{1,n} = \int_{(n-1)T}^{nT} f(\tau)\,\mathrm{d}\tau = Tf_{n-1} - \frac{T^2}{2} f'(\xi), \tag{4.60}$$

where $(n - 1)\,T < \xi < nT$. The second term in relation (4.60) determines the approximation error for the integral. We neglect this term whereby we obtain a sequence of approximate values of the integral for which the notation $\{i_{1,R,n}\}$ is introduced. Upon transformation of this sequence and substitution into (4.59) we obtain the following transform of the sequence of values of the approximate solution of the integral

$$G_{1,R}(z) = \frac{z}{z-1} Tz^{-1} F(z) = \frac{T}{z-1} F(z).$$

Thus, the transfer function of the discrete integrator, which consists in the application of the *rectangular rule*, is (with error proportional to interval T)

$$G_{1,R}(z) = \frac{T}{z-1}. \tag{4.61}$$

If the *trapezoidal rule* is used for the approximation of integral (4.58), with $m = 1$ again, we have

$$\int_{(n-1)T}^{nT} f(\tau)\,\mathrm{d}\tau = \frac{T}{2}(f_n + f_{n-1}) - \frac{T^3}{12} f''(\xi),$$

where ξ lies again in the interval $[(n-1)T, nT]$. By a procedure similar to the one discussed in the preceding case we obtain the transform of the sequence of values of the approximate solution of the integral in the form

$$H_{1,T}(z) = \frac{z}{z-1}\frac{T}{2}(1+z^{-1})\,F(z) = \frac{T}{2}\frac{z+1}{z-1}F(z)\,.$$

This yields the following transfer function of the discrete integrator (based on the trapezoidal rule with error proportional to T^2):

$$G_{1,T}(z) = \frac{T}{2}\frac{z+1}{z-1}\,. \tag{4.62}$$

The evaluation of integral (4.58) by the Newton–Cotes formulae for order $m > 1$ yields, upon application of the $\mathscr{Z}$ transform and modification, a set of further transfer functions of discrete integrators. Of these we present the following two:

$m = 2$ – *Simpson's rule* (error proportional to T^4):

$$G_2(z) = \frac{T}{3}\frac{z^2+4z+1}{z^2-1}\,; \tag{4.63}$$

$m = 4$ – *Boole's rule* (error proportional to T^6):

$$G_4(z) = \frac{2}{45}T\,\frac{7z^4+32z^3+12z^2+32z+7}{z^4-1}\,. \tag{4.64}$$

The properties of discrete integrators can be evaluated according to the deviation of their complex frequency characteristics from the complex frequency characteristic of the continuous integrator $G_L(j\omega) = 1/(j\omega)$. For the evaluation of this deviation we introduce the complex frequency *error function* by the relation

$$E_m(j\omega) = \frac{G_m(e^{j\omega T})}{G_L(j\omega)} = j\omega\, G_m(e^{j\omega T})\,. \tag{4.65}$$

In Fig. 44 the behaviour of the modules (absolute values) of this error function is presented in logarithmic scale for the mentioned types of discrete integrators in dependence on ωT. The transfer functions of the discrete integrators – with the exception of $G_{1,R}(z)$ – have symmetrical polynomials in the numerator as well as in the denominator. Therefore, their phase responses are identical with the phase response of the continuous integrator. It thus suffices to appraise them only from the standpoint of the amplitude response.

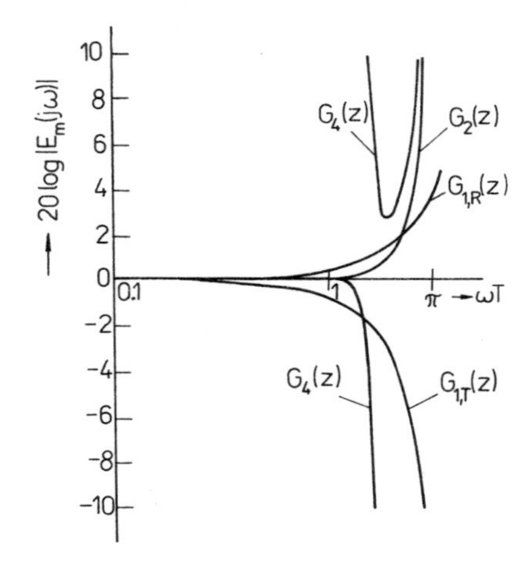

Fig. 44

The successful application of a discrete integrator instead of a continuous integrator assumes that the amplitude response of the error function be close to one till to the close neighbourhood of half the radian sampling frequency, i.e. to $\omega = \omega_s/2 = \pi/T$.

The comparison of the individual cases in Fig. 44 implies that the application of approximations of higher order does not always lead to better frequency properties of discrete integrators. Thus, for instance, Boole's rule which has approximation error proportional to T^6 is not suitable for the integration of signals containing components of frequencies in the neighbourhood of $T\omega = 1.5$.

Similar procedures could be used to represent quadrature formulae for equidistant intervals presented in the literature covering the field of numerical calculus. We would obtain transfer functions of discrete integrators from whose amplitude responses we could assess the adequateness of their application in various applications.

In the next part of this subsection we indicate how to construct discrete integrators exploiting the principle of the invariance of the response to a signal of the $t^{\nu-1}$ type, discussed in detail in Subsection 4.1.1. Applying this method to the transfer function $G_L(p) = 1/p$, we obtain for the transfer function of the discrete integrator which is invariant to the signal $f(t) = t^{\nu-1}/(\nu - 1)!$ the following formula

$$G_\nu(z) = \frac{\mathscr{Z}\left\{\mathscr{L}^{-1}\left\{\dfrac{1}{p^{\nu+1}}\right\}\Big|_{t=nT}\right\}}{\mathscr{Z}\left\{\mathscr{L}^{-1}\left\{\dfrac{1}{p^{\nu}}\right\}\Big|_{t=n/T}\right\}}. \tag{4.66}$$

From the list of transforms in Section 7.2 we determine the transfer functions of discrete integrators for $\nu = 1, 2, 3, 4$.

Invariance to the unit step, $\nu = 1$:

$$G_1(z) = \frac{T}{z - 1}.$$

The obtained function is identical with the transfer function corresponding to the rectangular approximation.

Invariance to the function $f(t) = t$, $\nu = 2$:

$$G_2(z) = \frac{T}{2}\,\frac{z + 1}{z - 1}.$$

The transfer function corresponds to integration according to the trapezoidal rule.

Invariance to the function $f(t) = t^2/2$, $\nu = 3$:

$$G_3(z) = \frac{T}{3}\,\frac{z^2 + 4z + 1}{(z - 1)(z + 1)}.$$

The transfer function corresponds to integration according to the Simpson rule.

Invariance to the function $f(t) = t^3/3!$, $\nu = 4$:

$$G_4(z) = \frac{T}{4}\,\frac{(z + 1)(z^2 + 10z + 1)}{(z - 1)(z^2 + 4z + 1)}.$$

Discrete integrators for $\nu \geqq 4$ are unstable in the sense of the one-sided $\mathscr{Z}$ transform. When using them we thus have to proceed in line with Subsection 2.4.2 which deals with the inverse two-sided $\mathscr{Z}$ transform.

With the exception of $G_1(z)$, the phase responses of all the transfer functions of discrete integrators are identical with the phase response of the continuous integrator. Therefore, it is possible to assess the properties of discrete integrators only by the module of the error function $E_m(j\omega)$ in accordance with relation (4.65). Fig. 45 shows, in logarithmic scale, the frequency behaviours of the modules of the error function in dependence on ωT.

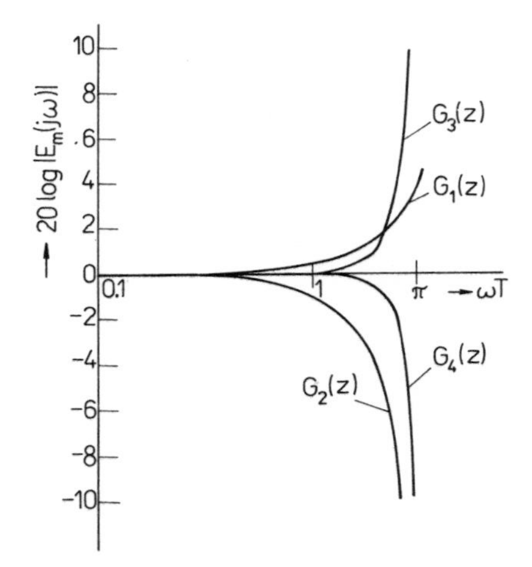

Fig. 45

The last method of the construction of transfer functions of discrete integrators discussed is the procedure which makes use of the expansion of the function $\ln z$ into the continued fraction (4.48). This was used in the beginning of Subsection 4.2.2 when deriving the bilinear transformation.

Let us substitute for $p = T^{-1} \ln z$ in the transfer function of the continuous integrator. We obtain the relation

$$G_L(p) = \frac{1}{p} = \frac{T}{\ln z},$$

in which $\ln z$ can be replaced by its infinite expansion into a continued fraction. If we limit ourselves to a finite number of terms, we obtain a sequence of transfer functions of discrete integrators which approximate the frequency properties of the continuous integrator in the frequency band $[0, \pi/T]$ as follows:

$$G_1(z) = \frac{T}{2}\,\frac{z+1}{z-1},$$

$$G_2(z) = \frac{T}{3}\,\frac{z^2+4z+1}{(z-1)(z+1)},$$

$$G_3(z) = 3T\,\frac{(z+1)(z^2+8z+1)}{(z-1)(11z^2+38z+11)},$$

$$G_4(z) = \frac{6}{5}\,T\,\frac{z^4+16z^3+36z^2+16z+1}{(z-1)(z+1)(5z^2+32z+5)};$$

generally, for $k \geqq 2$, we have

$$G_k(z) = T\,\frac{(2k-1)(z+1)\,Q_{k-1} - (k-1)^2(z-1)^2\,Q_{k-2}}{(2k-1)(z+1)\,P_{k-1} - (k-1)^2(z-1)^2\,P_{k-2}}, \quad (4.67)$$

where

$$Q_0 = 1, \qquad P_0 = 1,$$
$$Q_1 = z+1, \qquad P_1 = 2(z-1).$$

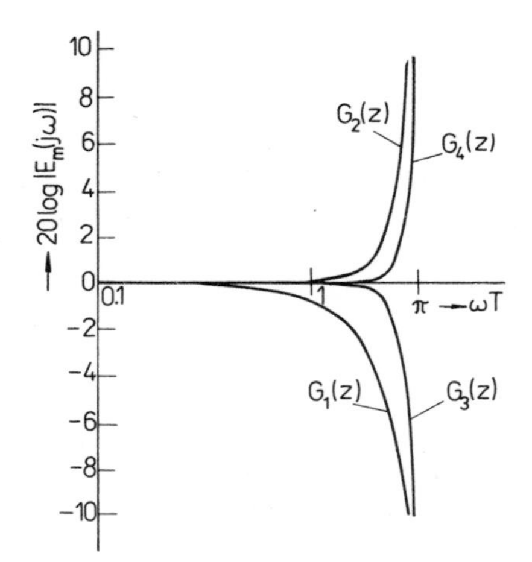

Fig. 46

In the first two discrete transfer functions we recognize the transfer functions of the discrete integrators which correspond to the application of the trapezoidal and Simpson's rules. The polynomials in the numerators as well as the denominators of all the functions are symmetrical. Thus, all the discrete integrators have phase response identical with the phase response of the continuous integrator. However, symmetrical polynomials are also a symptom of the fact that starting with the function $G_3(z)$ some of the poles lie outside the unit circle. Thus, the discrete integrators are unstable in the sense of the one-sided $\mathscr{Z}$ transform and they have to be used in the sense of Subsection 2.4.2.

The behaviour of the modules of the error function defined by relation (4.65) is indicated in Fig. 46, in the logarithmic scale, for the individual transfer functions as functions of ωT. Comparison of the frequency responses of all the discussed discrete integrators shows that best properties can be achieved by the application of transfer functions derived from the expansion of the function $\ln z$ into a continued fraction. For the prescribed error in the frequency domain the degree of these functions (for $k > 2$) is lower than the degree of the transfer functions which satisfy the given requirements with the aid of the other discussed approximation methods.

4.3.2. *Numerical Differentiation*

In the theory of discrete signals and systems we often encounter the problem of determining the values of the derivative of a continuous function $f(t)$ if this function is given only by a sequence of its values for $t = nT$, $n = 0, 1, 2, \ldots, T > 0$. As a rule, this problem is solved by fitting an interpolation polynomial $h(t)$ of the continuous variable t to the sequence of values $f(nT)$; this polynomial is then differentiated. Sampling the differentiated polynomial $h'(t)$ for $t = nT$ yields a sequence of values $h'_n = h'(nT)$. For the construction of the interpolation polynomial we may use, e.g., the Newton, Bessel, or Stirling interpolation formulas.

For the design of discrete differentiators which simulate the behaviour of continuous differentiators, the Stirling interpolation formula proves best. It is based on the application of central differences and it guarantees higher accuracy of the obtained approximation of the derivative.

Let us assume that the function $f(t)$ is given by a sequence of its values $f(nT)$. Applying the *Stirling formula* we obtain, for the numerical differentiation of the function $f(t)$ at the points $t = nT$, the relation

$$h_n = \left.\frac{\mathrm{d}f(t)}{\mathrm{d}t}\right|_{t=nT} = \frac{1}{2T}\left[(\Delta f_n + \Delta f_{n-1}) - \right.$$
$$\left. - \frac{1}{6}(\Delta^3 f_{n-1} + \Delta^3 f_{n-2}) + \frac{1}{30}(\Delta^5 f_{n-2} + \Delta^5 f_{n-3}) - \dots \right]. \tag{4.68}$$

According to the number of differences used we obtain, for the sequence of approximate values of the derivatives, a set of equations which differ in the accuracy of the approximation. Let us represent the differences in these relations by the sequence $\{f_n\}$; thus, we obtain

$$h_{1,n} = \frac{1}{2T}(f_{n+1} - f_{n-1}),$$

$$h_{2,n} = \frac{1}{12T}(-f_{n+2} + 8f_{n+1} - 8f_{n-1} + f_{n-2}),$$

$$h_{3,n} = \frac{1}{60T}(f_{n+3} - 9f_{n+2} + 45f_{n+1} - 45f_{n-1} + 9f_{n-2} - f_{n-3}).$$

The application of the $\mathscr{Z}$ transform yields the transform of the sequences of values of the derivatives for which we introduce the notation $\mathscr{Z}\{h_{k,n}\} = H_k(z)$ for $k = 1, 2, 3$:

$$H_1(z) = \frac{1}{2T}\left[(z - z^{-1}) F(z) - zf_0\right],$$

$$H_2(z) = \frac{1}{12T}\left[(-z^2 + 8z - 8z^{-1} + z^{-2}) F(z) + \right.$$
$$\left. + f_0(z^2 - 8z) + zf_1\right],$$

$$H_3(z) = \frac{1}{60T}\left[(z^3 - 9z^2 + 45z - 45z^{-1} + 9z^{-2} - z^{-3}) F(z) - \right.$$
$$\left. - f_0(z^3 - 9z^2 + 45z) - f_1(z^2 - 9z) - f_2 z\right].$$

If we put, in the transforms $H_k(z)$, the initial conditions equal to zero, it is possible to define the transfer functions of the discrete differentiators based on the Stirling interpolation formula by the equations

$$G_1(z) = \frac{H_1(z)}{F(z)} = \frac{1}{2T}\,\frac{z^2 - 1}{z}, \tag{4.69}$$

$$G_2(z) = \frac{H_2(z)}{F(z)} = \frac{1}{12T}\,\frac{-z^4 + 8z^3 - 8z + 1}{z^2}, \tag{4.70}$$

$$G_3(z) = \frac{H_3(z)}{F(z)} = \frac{1}{60T}\,\frac{z^6 - 9z^5 + 45z^4 - 45z^2 + 9z - 1}{z^3}. \tag{4.71}$$

The obtained transfer functions correspond to noncausal discrete systems. All the transfer functions have a pole of multiplicity k at the point $z = 0$. Thus, they can be modified upon multiplication by the factor z^{-k} to transforms in the sense of the one-sided $\mathscr{Z}$ transform. These can then be interpreted as transforms of delayed impulse responses of causal discrete differentiators.

The described approach to the construction of discrete differentiators can be viewed as the simulation of continuous differentiators in the time domain. Now, let us indicate the approach to the construction of transfer functions of discrete differentiators which corresponds to simulation in the frequency domain.

The transfer function of a continuous differentiator is the reciprocal transfer function of a continuous integrator. In principle, we could thus obtain transfer functions of discrete differentiators from reciprocal transfer functions of discrete integrators derived in Subsection 4.3.1. However, we would arrive at transfer functions with infinite impulse response which would, moreover, be mostly unstable. Therefore, we shall use the general method for the expansion of the function $\ln z$ indicated by N. Obreschkoff [18].

For the transfer function of a continuous differentiator we then have

$$G_L(p) = p = \frac{\ln z}{T}. \tag{4.72}$$

The function $\ln z$ can be approximated by the series

$$\ln z = \sum_{i=1}^{k} \frac{\binom{k}{i}}{i\binom{2k}{i}} \left[(-1)^{i-1} + z^{-i}\right](z-1)^i .$$

Upon substitution into (4.72) we obtain the general relation

$$G_L(p) \doteq G_k(z) = -\frac{z-1}{T} \frac{\sum_{i=1}^{k} \frac{\binom{k}{i}}{\binom{2k}{i}} z^{k-i}(z-1)^{i-1}\left[(-z)^i - 1\right]}{z^k} \tag{4.74}$$

which is the transfer function of a discrete differentiator of order k. Substitution for k yields

$$G_1(z) = \frac{1}{2T}\frac{z^2-1}{z} = \frac{1}{2T}\frac{(z-1)(z+1)}{z},$$

$$G_2(z) = \frac{1}{12T}\frac{-z^4+8z^3-8z+1}{z^2} = -\frac{1}{12T}\frac{(z-1)(z+1)(z^2-8z+1)}{z^2},$$

$$G_3(z) = \frac{1}{60T}\frac{z^6-9z^5+45z^4-45z^2+9z-1}{z^3} = \frac{1}{60T}\frac{(z-1)(z+1)(z^4-9z^3+46z^2-9z+1)}{z^3},$$

$$G_4(z) = \frac{1}{810T}\frac{-3z^8+32z^7-168z^6+672z^5-672z^3+168z^2-32z+3}{z^4} = -\frac{1}{810T}\frac{(z-1)(z+1)(3z^6-32z^5+171z^4-704z^3+171z^2-32z+3)}{z^4}.$$

We have arrived at the interesting result that the transfer functions of discrete differentiators constructed in the frequency domain are identical with transfer functions constructed in the time domain based on the Stirling interpolation formula.

When evaluating the properties of the obtained discrete differentiators in the frequency domain we can limit ourselves to the assessment of the behaviour of the module of the complex error function since the phase characteristics of all the corresponding transfer functions are identical with the phase characteristics of the continuous differentiator.

$$E_k(\mathrm{j}\omega) = \frac{G_k(\mathrm{e}^{\mathrm{j}\omega T})}{G_\mathrm{L}(\mathrm{j}\omega)} = \frac{G_k(\mathrm{e}^{\mathrm{j}\omega T})}{\mathrm{j}\omega}. \tag{4.75}$$

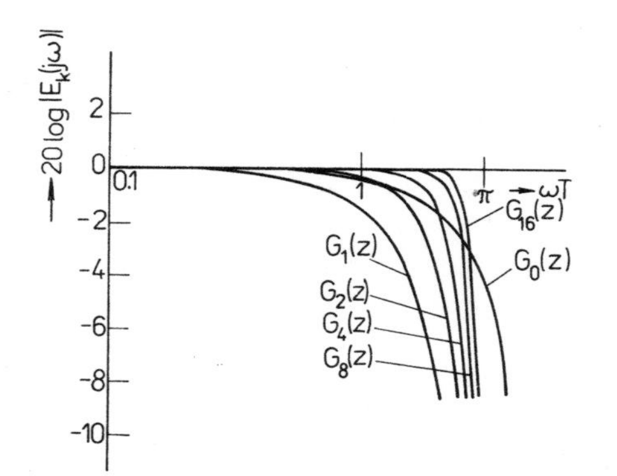

Fig. 47

In Fig. 47 the behaviour of the module $|E_k(\mathrm{j}\omega)|$ is shown in logarithmic scale in dependence on ωT for $k = 1, 2, 4, 8, 16$. For comparison, the module of the error function is also indicated for the transfer function of the discrete differentiator

$$G_0(z) = \frac{1}{T}(z - 1).$$

This corresponds to the replacement of the derivative by the first order difference. (However, the phase response is not ideal for $G_0(z)$.)

The problem of the construction of the transfer function of a discrete differentiator can still be treated in line with other – entirely different – approaches which consist in the optimization of the parameters by the linear programming method. However, this approach is related more or less to approximation for the synthesis of frequency selective filters which is discussed, e.g., in [2], [22], [29], [32].

Chapter 5

Application of the $\mathscr{Z}$ Transform to the Analysis of Digital Filters

5.1 Realization of the Transfer Function of a Digital Filter

In Section 3.2 some of the basic concepts from the theory of linear discrete systems were explained. Now, we turn our attention to a special case of discrete systems, namely to digital filters which were mentioned briefly in Section 1.1.

A digital filter is most frequently realized by a device which accomplishes a required algorithm of discrete signal processing, with limited accuracy, as a rule in fixed point arithmetic. As an example we mention the implementation of a frequency selective filter by a microprocessor or by a signal processor. In such cases it is necessary to choose and analyze carefully the algorithm for the processing of the given signal so that the digital filter satisfy the imposed requirements as far as speed as well as accuracy are concerned. In brief, it is possible to summarize that the *problems solved by means of digital filters are problems of circuit and signal theory solved by methods of numerical calculus, with the aid of computer technology.*

In this chapter we attempt to build up the foundations of the *analysis of digital filters* determined not only by their transfer function but also by their structure which is given by the interconnection of several basic blocks. Effective and rapid analysis depends on the manner of description of the network which has to enable not only the computation of the response of a digital filter to given input signals but also the computation of the internal signals of the filter. With the aid of these signals it is then possible to evaluate, e.g., the dynamic range of the filter, the influence of signal quantization, etc. Since the objects of the analysis are not only the time behaviour of signals but also their spectral properties, the analysis has to enable the transition from the time to the frequency domain. Moreover, the manner of description must be general to the extent that it is applicable for different structures, thus making possible their comparison.

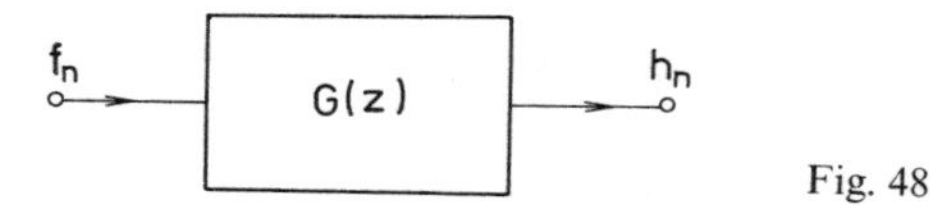

Fig. 48

The *realization of a digital filter* consists in the formulation of the algorithm for the computation of its response to an input signal if the transfer function $G(z)$ is given. For the transform of its response $H(z) = \mathscr{Z}\{h_n\}$, see Fig. 48, we have

$$H(z) = F(z)\, G(z)\,, \tag{5.1}$$

where $F(z) = \mathscr{Z}\{f_n\}$ is the transform of the input signal $\{f_n\} = \{f_0, f_1, f_2, \ldots\}$.

Assume now that $G(z)$ is the transfer function of a causal digital filter with infinite impulse response (IIR)

$$G(z) = \frac{a_0 z^s + a_1 z^{s-1} + \ldots + a_s}{z^s + b_1 z^{s-1} + \ldots + b_s} = \frac{P(z)}{Q(z)}. \tag{5.2}$$

This relation differs from relation (3.44) in that the order of the numerator is equal to the order of the denominator and $b_0 = 1$; this does not affect the generality of the considerations. Now, let us substitute for $G(z)$ into (5.1) and multiply both sides by the denominator $Q(z)$. To be able to apply the theorem on the translation of a sequence, with regard to causality it is namely for right translation, i.e. for delay, we divide both sides by the quantity z^s. Thus, we obtain the implicit relation

$$H(z) = F(z) \sum_{i=0}^{s} a_i z^{-i} - H(z) \sum_{i=1}^{s} b_i z^{-i} = \tag{5.3a}$$

$$= a_0\, F(z) + \sum_{i=1}^{s} \left[a_i\, F(z) - b_i\, H(z)\right] z^{-i}\,. \tag{5.3b}$$

By execution of the inverse transform we obtain recurrent relations for the response sequence $\{h_n\}$

$$h_n = \sum_{i=0}^{s} a_i f_{n-1} - \sum_{i=1}^{s} b_i h_{n-i} = \tag{5.4a}$$

$$= a_0 f_n + \sum_{i=1}^{s} (a_i f_{n-i} - b_i h_{n-i})\,. \tag{5.4b}$$

We have arrived at relations which are identical with the result of Example 2 in Subsection 2.2.2. There the so-called strip method for the inverse $\mathscr{Z}$ transform of the product of transforms was introduced, with one of the transforms in the form of a series in the variable z^{-1} and the other in the form of a rational function. The recurrent relation (5.4a), and thus also the corresponding strip algorithm, can be graphically represented by the network shown in Fig. 49. The following three types of blocks necessary for the evaluation of relation (5.4a) are included in the network:

- multiplication of a sequence by a constant,
- addition of two sequences,
- delay of a sequence by one sampling interval, i.e. by time $T = 1/f_s$.

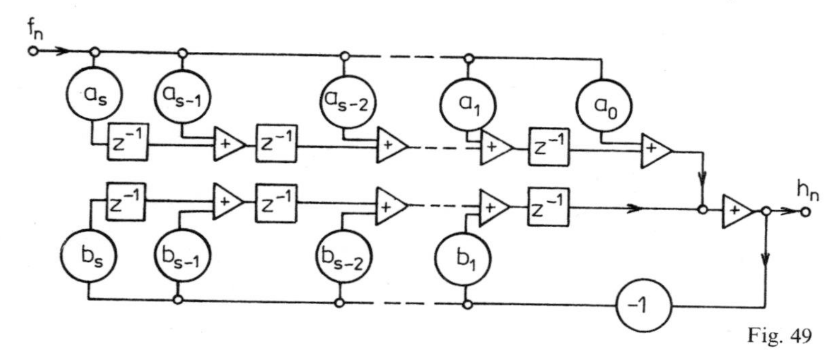

Fig. 49

From Fig. 49 we see that it is possible to realize a digital filter by applying a more universal *functional block* shown in Fig. 50a. This functional block includes two multipliers, a two-input adder, and a block delaying the output sequence of the adder by one delay interval. For the ith block shown in Fig. 50a we have, in the time domain,

$$v_{1,n}^{(i)} = a_1^{(i)} u_{1,n}^{(i)} + a_2^{(i)} u_{2,n}^{(i)},$$
$$v_{2,n}^{(i)} = v_{1,n-1}^{(i)}.$$

a) b) Fig. 50

For the $\mathscr{Z}$ transform we obtain

$$V_1^{(i)} = a_1^{(i)} U_1^{(i)} + a_2^{(i)} U_2^{(i)},$$
$$V_2^{(i)} = z^{-1} V_1^{(i)},$$

where, e.g., $V_1^{(i)} = V_1^{(i)}(z) = \mathscr{Z}\{v_{1,n}^{(i)}\}$. To abbreviate the notation, the functional dependence of the transform on the variable z will be omitted in what follows.

The basic functional block is represented by a *directed graph* in accordance with Fig. 50b. The input, internal, and output quantities are represented in the graph by circles which are called *nodes.* To each node the transform of the corresponding quantity is written. Relations between quantities are represented by directed line segments which are called *branches.* Each branch is assigned transmission which is equal to the coefficient of its input quantity. In the case of input branches we write the transmission to the *full arrowhead*, e.g. $a_1^{(i)} = V_1^{(i)}/U_1^{(i)}$ $(U_2^{(i)} = 0)$, while for the output branch the transmission $z^{-1} = V_2^{(i)}/V_1^{(i)}$ is not included and the respective *arrowhead is left empty.* If $a_1^{(i)} = 1$, or if $a_2^{(i)} = 1$, the legend is also not included. Consequently, we come to know at first sight which branches modify the input signal, which are connecting branches, and which are delay branches.

A node with no input branch and with at least one output branch is called a *source node* and its quantity is denoted by $F = F(z)$. On the other hand, a node with input branches only is called a *sink node* and it is denoted by $H = H(z)$. The remaining nodes are *internal nodes* of the graph. The ratio of the quantity of the sink H and the quantity of the source F is the *transmission of the graph* G. It is equal to the transfer function of a digital filter with one input and one output. Thus, we have $G = G(z) = \mathscr{Z}\{g_n\} = H/F$. The ratio of the quantity $V_1^{(k)}$ and the quantity $V_1^{(i)}$ is the transmission of the subgraph and it is denoted by $G_{ik} = V_1^{(k)}/V_1^{(i)}$.

It is obvious that *degenerated blocks* also exist if, e.g., $a_1^{(i)} = 0$ or $a_1^{(i)} = 1$, or if the delay element is absent. In Fig. 51 some of the used degenerations of the basic block are summarized.

Applying the described formalism we can draw a graph (see Fig. 52) which corresponds to the implementation of relation (5.4a). So far, the internal quantities were not denoted for the sake of simplicity; the coefficients of the transfer function are entered to the full arrowheads.

Graph	Application	Function
 $a_1^{(i)}$ $U_1^{(i)}$ $V_1^{(i)}$	scaling	$V_1^{(i)} = a_1^{(i)} U_1^{(i)}$
 $U_1^{(i)}$ $V_1^{(i)}$	connecting branch	$V_1^{(i)} = U_1^{(i)}$
$U_1^{(i)}$ $V_1^{(i)}$	disconnected branch	$V_1^{(i)} = 0 \,.\, U_1^{(i)}$
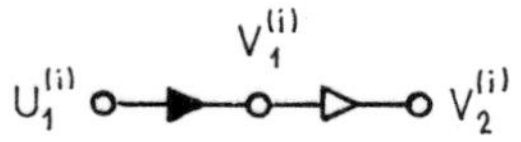 $V_1^{(i)}$ $U_1^{(i)}$ $V_2^{(i)}$	delay branch	$V_2^{(i)} = z^{-1} U^{(i)}$
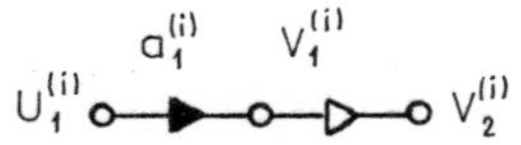 $a_1^{(i)}$ $V_1^{(i)}$ $U_1^{(i)}$ $V_2^{(i)}$	scaling with delay	$V_2^{(i)} = a_1^{(i)} z^{-1} U_1^{(i)}$
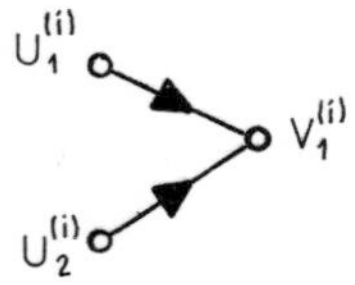 $U_1^{(i)}$ $V_1^{(i)}$ $U_2^{(i)}$	sum	$V_1^{(i)} = U_1^{(i)} + U_2^{(i)}$
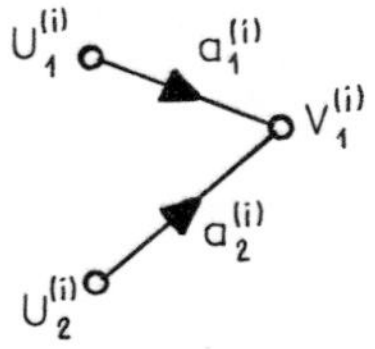 $U_1^{(i)}$ $a_1^{(i)}$ $V_1^{(i)}$ $a_2^{(i)}$ $U_2^{(i)}$	weighted sum	$V_1^{(i)} = a_1^{(i)} U_1^{(i)} + a_2^{(i)} U_2^{(i)}$

Fig. 51

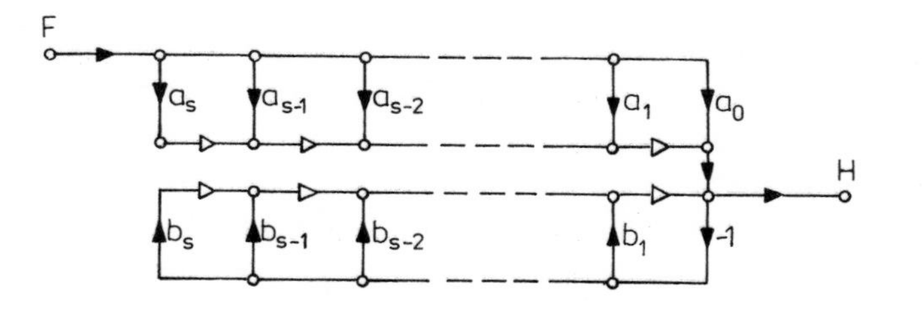

Fig. 52

The signal-flow graph is a useful and clear graphical representation of the dependences between given and desired quantities in a digital filter. In Section 5.2 we shall show how a digital filter can be described by a matrix and a table with the aid of which it is possible to determine the filter's time response.

5.1.1. *Digital Filters with Infinite Impulse Response*

Above we have tried to indicate the approach to the realization of a transfer function in the form of a ratio of two polynomials. In other words, this corresponds to finding the algorithm for the computation of the response of the corresponding digital filter. Now, we shall direct our attention to some special cases of realization and their characteristic properties.

As mentioned in Section 3.2, to a transfer function of the form (5.2) an impulse response corresponds which is given as a linear combination of exponential sequences, thus being infinite. Digital filters with this property are, therefore, called filters with *infinite impulse response* – IIR filters for short. For a transfer function to be of this type the polynomial $Q(z)$ must have at least one nonzero coefficient (naturally, we assume the polynomials have no common divisor). A consequence thereof is the recurrent character of relation (5.4), manifested by *feedback* in the lower part of Fig. 52. For this reason the discussed realization is called, in the literature, the *recursive realization* of the transfer function.

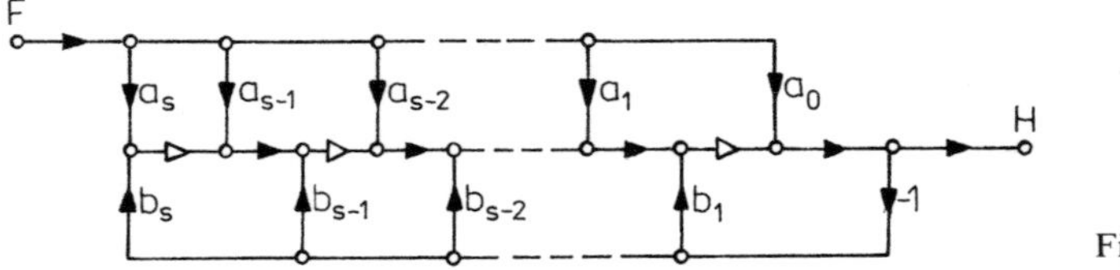

Fig. 53

If we start from relation (5.3b) when realizing the transfer function $G(z)$, we obtain a further structure which is shown in Fig. 53. This structure has a minimal number of adder, multiplier, and delay blocks. In the literature this structure is usually called the *1st canonic form* of the network realizing the transfer function (5.2). It is obvious that the structure is recursive and of the IIR type.

Another basic structure which is a realization of the transfer function (5.2) is the *2nd canonic form* shown in Fig. 54. It is obtained either by the realization of a transfer function decomposed into two parts

$$G(z) = (a_0 + a_1 z^{-1} + \dots + a_s z^{-s}) \frac{1}{1 + b_1 z^{-1} + \dots + b_s z^{-s}} \tag{5.5}$$

or by the transposition of the 1st canonic form [29]. *Transposition* means reversing all the arrowheads in the figure and interchanging the input and output.

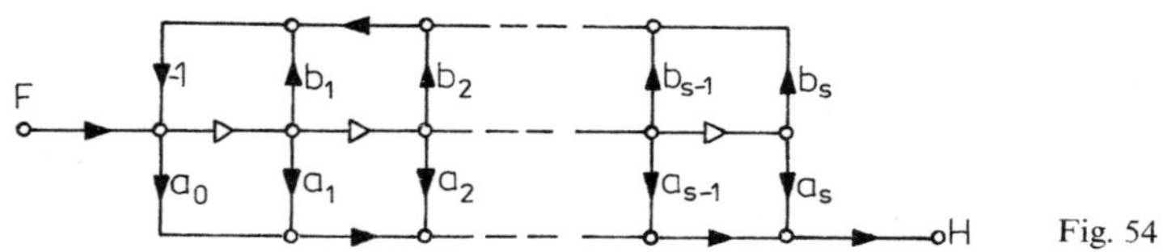

Fig. 54

Upon decomposing the numerator and denominator into root factors we can write transfer function $G(z)$ in the form of a product of simpler transfer functions

F G_{C1} G_{Cs} H

Fig. 55

$$G(z) = \prod_{i=1}^{s} G_{Ci}(z)\,. \tag{5.6}$$

This leads to the cascade connection of simpler digital filters in line with Fig. 55 where the empty arrowheads stand for the transfer functions of the partial filters $G_{Ci} = G_{Ci}(z)$. As a rule, this structure is called the *3rd canonic form.* The individual transfer functions $G_{Ci}(z)$ are realized, as a rule, in the 1st and 2nd canonic forms. With regard to the simplicity of the realization it is reasonable to associate complex conjugate zeros and poles in the partial transfer functions since this guarantees real coefficients in the network.

The last basic structure of a recursive filter which we present here consists in the decomposition of a transfer function into partial fractions

$$G(z) = \sum_{i=1}^{s} G_{Pi}(z)\,. \tag{5.7}$$

It is given by parallel combination of partial filters, see Fig. 56. As a rule, this structure is called the *4th canonic form.* The partial transfer functions are of the first and second orders and they are realized in the 1st or 2nd canonic forms again.

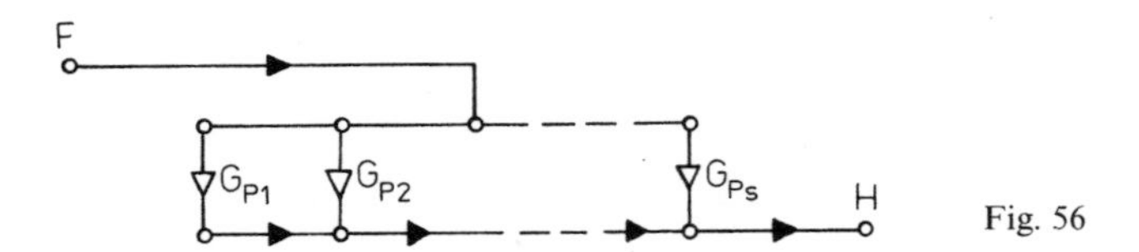

Fig. 56

Besides the classical canonic forms there exist numerous further canonic as well as noncanonic structures of filters with infinite impulse response which are obtained, e.g., by equivalent transformations of the network, or they are based on some special decomposition of the transfer function. Ladder, lattice, and wave digital filters are examples of such structures [2], [25].

5.1.2. *Digital Filters with Finite Impulse Response*

Now, let us assume that in relation (5.2) all the coefficients in the denominator of transfer function $G(z)$ vanish, i.e. $b_i = 0$ for $i = 1, 2, \ldots, s$. Then the transfer function is given by a polynomial of degree s in the variable z^{-1}. As a rule, we choose $s = N - 1$, where N is the number of coefficients of the filter. The transfer function of a digital filter with finite impulse response is

$$G(z) = \sum_{i=0}^{N-1} a_i z^{-i}. \tag{5.8}$$

The inverse transform yields impulse response with only N terms. Therefore, this type of filter is called *filter with finite impulse response* – FIR filter for short.
The transform of the response is in this case given by

$$H(z) = F(z) \sum_{i=0}^{N-1} a_i z^{-i} = \sum_{i=0}^{N-1} a_i F(z) z^{-i}. \tag{5.9}$$

Hence, the inverse transform yields the response

$$h_n = \sum_{i=0}^{N-1} a_i f_{n-i} \tag{5.10}$$

which can be viewed as a weighted sum of a finite number of elements of the input sequence $\{f_n\}$.

Although it is possible to realize this transfer function in all the four basic canonic forms we present only the two which are used most, i.e. the *1st and 2nd canonic forms.* The two structures follow from Figs. 52 and 54 in which we put $b_i = 0$. These filters do not include feedback which is the consequence of the recurrent computation of the response. For this reason they are also called *nonrecursive.* They are always stable.

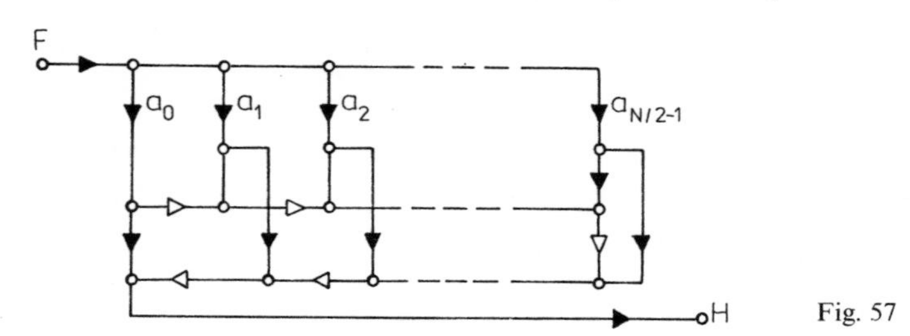

Fig. 57

An important group of FIR filters are filters which have *symmetric* or *antisymmetric transfer functions*, i.e. for which $a_i = a_{N-i-1}$ or $a_i = -a_{N-i-1}$ holds, respectively. In the case of symmetric coefficients we obtain for the 1st canonic form and even N the network of Fig. 57, for the 2nd canonic form and even N the network of Fig. 58. In the next section we shall show that the transfer functions of these filters have zeros either on the unit circle or symmetrically placed with respect to the unit circle, thus having linear phase response.

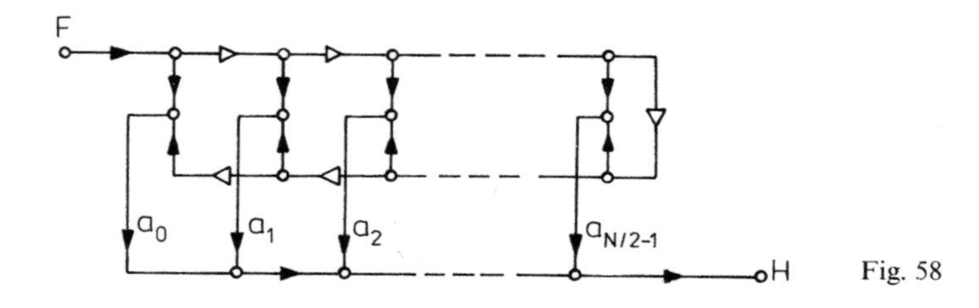

Fig. 58

As a special case of the realization of a filter with finite impulse response we present its *recursive realization* which consists in the following decomposition of its transfer function into partial fractions:

$$G(z) = \sum_{i=0}^{N-1} a_i z^{-i} = \frac{z^N - 1}{N z^N} N \frac{\sum_{i=0}^{N-1} a_i z^{N-1-i}}{z^N - 1} =$$

$$= \frac{1 - z^{-N}}{N} \sum_{i=0}^{N-1} \frac{z\, G\left(e^{j(2\pi/N)i}\right)}{z - e^{j(2\pi/T)i}}. \tag{5.11}$$

The expression before the summation sign is the transform of the relative difference over N samples. The individual partial transfer functions possess poles equidistantly distributed along the unit circle and their coefficients are given by the values of the transfer function $G(z)$ for $z = e^{j(2\pi/N)i}$, $i = 0, 1, \ldots, (N-1)$. For this reason the resulting structure of Fig. 59 is also called *frequency-sampling realization*. Since the transfer function has poles on the unit circle it is unstable. For this function to be realizable we have to shift the poles slightly into the unit circle.

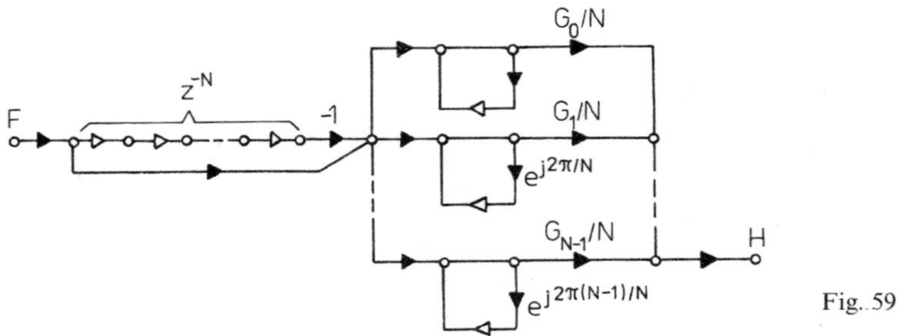

Fig. 59

5.1.3. *Computation of Frequency Characteristics of Digital Filters*

In Subsection 3.2.2 frequency characteristics of linear discrete systems were introduced and some of their basic properties were given. Now, let us discuss several algorithms of their computation for the individual types of transfer functions.

The complex frequency response of a digital filter is given by its transfer function for $z = e^{j\omega T} = \cos \omega T + j \sin \omega T$. In all transfer functions we encounter polynomials, eventually expressed by root factors. It is therefore reasonable to devote our attention to algorithms for the computation of the functional values of a polynomial and of its derivative which are economical as far as the number of arithmetic operations is concerned. We shall give two algorithms which consist in the division of a polynomial by a polynomial and are in essence special cases of the strip method of Subsection 2.2.2.

The first algorithm is the so-called *Horner scheme* which is of advantage only for the computation of the value of a polynomial with complex coefficients, or for the computation of a polynomial with real coefficients but at a real point. The second and less familiar algorithm, the *Collatz scheme* for the computation of the value of a polynomial with real coefficients at a complex point, works altogether with real numbers. It can be also applied to the computation for a pair of real numbers. Assume a polynomial in the form

$$P(z) = a_0 z^s + a_1 z^{s-1} + \dots + a_s$$

and the variable $z = u + \mathrm{j}v$.

(a) *Horner scheme.*

We have

$$\frac{P(z)}{z - z_0} = R(z) + \frac{A}{z - z_0},$$

$$\frac{R(z)}{z - z_0} = S(z) + \frac{B}{z - z_0},$$

whence

$$P(z) = (z - z_0)\, R(z) + A\,,$$

$$R(z) = (z - z_0)\, S(z) + B\,.$$

For $z = z_0$ we then obtain

$$P(z_0) = A\,,$$

and since

$$\frac{\mathrm{d}\,P(z)}{\mathrm{d}z} = R(z) + R'(z)\,(z - z_0)\,, \qquad R'(z) = \frac{\mathrm{d}\,R(z)}{\mathrm{d}z}\,,$$

we have

$$\frac{\mathrm{d}\,P(z)}{\mathrm{d}z} = R(z_0) = B\,.$$

The computations can be organized according to the following scheme:

Polynomial	z_0	z_0
a_0	$a'_0 = a_0$	$a''_0 = a'_0$
a_1	$a'_1 = a_1 - z_0 a'_0$	$a''_1 = a'_1 - z_0 a''_0$
a_2	$a'_2 = a_2 - z_0 a'_1$	$a''_2 = a'_2 - z_0 a''_1$
$\vdots$	$\vdots$	$\vdots$
		a''_{s-2}
a_{s-1}	a'_{s-1}	$B = a'_{s-1} - z_0 a''_{s-2}$
a_s	$A = a_s - z_0 a'_{s-1}$	

(b) *Collatz scheme.*

We have

$$\frac{P(z)}{z^2 + \alpha z + \beta} = R(z) + \frac{Az + B}{z^2 + \alpha z + \beta},$$

$$\frac{R(z)}{z^2 + \alpha z + \beta} = S(z) + \frac{Cz + D}{z^2 + \alpha z + \beta},$$

where

$$z^2 + \alpha z + \beta = (z - z_0)(z - z_0^*) = z^2 - 2u_0 z + u_0^2 + v_0^2 .$$

Similarly as in the case of the Horner scheme we obtain

$$P(z_0) = Az_0 + B = (Au_0 + B) + \mathrm{j}Av_0 ,$$

$$\frac{\mathrm{d}\, P(z)}{\mathrm{d}z} = az_0 + b ,$$

where

$$a = 2D - \alpha C , \qquad b = A - 2C\beta + D\alpha .$$

The computation is organized by the iterated strip method:

Polynomial			
a_0	$a'_0 = a_0$	$a''_0 = a'_0$	$-\beta$
a_1	$a'_1 = a_1 - \alpha a'_0$	$a''_1 = a'_1 - \alpha a''_0$	$-\alpha$
a_2	$a'_2 = a_2 - \alpha a'_1 - \beta a'_0$	$a''_2 = a'_2 - \alpha a''_1 - \beta a''_0$	$\longleftarrow$
$\vdots$	$\vdots$	$\vdots$	
		a''_{s-5}	Strip
		a''_{s-4}	of
a_{s-3}	a'_{s-3}	$C = a'_{s-3} - \alpha a''_{s-4} - \beta a''_{s-5}$	paper
a_{s-2}	a'_{s-2}	$D = a'_{s-2} - \beta a''_{s-4}$	
a_{s-1}	$A = a_{s-1} - \alpha a'_{s-2} - \beta a'_{s-3}$		
a_s	$B = a_s - \beta a'_{s-2}$		

It has to be stressed that when computing the values of B and D we do not multiply by the coefficient $(-\alpha)$.

For the computation of the complex frequency responses for $z = e^{j\omega T} = \cos \omega T + j \sin T$, i.e. for points lying on the unit circle, we have $\alpha = 2 \cos \omega T$ and $\beta = 1$. If the transfer function is in the form of a ratio of two polynomials, we use the respective scheme twice for every point of the frequency response. In the case of a transfer function of a filter with *finite impulse response* we have

$$G(z) = \sum_{n=0}^{N-1} a_n z^{-n},$$

and the complex frequency response is obtained by the direct application of the Collatz scheme for $z = \cos \omega T + j \sin \omega T$. However, if the impulse response is *symmetrical*, i.e. if $a_n = a_{N-n-1}$, it is possible to reduce the number of arithmetic operations necessary to obtain one value by almost one half. Exploiting symmetry we obtain for N *even*

$$G(z) = \sum_{n=0}^{N/2-1} a_n(z^{-n} + z^{-(N-n-1)}), \tag{5.12}$$

and for N *odd*

$$G(z) = \sum_{n=0}^{(N-1)/2-1} a_n(z^{-n} + z^{-(N-n-1)}) + A_{(N-1)/2}z^{-(N-1)/2}. \tag{5.13}$$

If we put $z = e^{j\omega T}$ again, we obtain for the complex frequency response, for N *even*

$$G(e^{j\omega T}) = e^{-j\omega T(N-1)/2} \,.\, 2\,\mathrm{Re}\left[z^{1/2} \sum_{n=0}^{N/2-1} a_n z^{-(n-N/2+1)} \right]\Bigg|_{z=e^{j\omega T}} = \tag{5.14a}$$

$$= e^{-j\omega T(N-1)/2} \,.\, 2 \sum_{n=0}^{N/2-1} a_n \cos\left(n - \frac{N-1}{2}\right)\omega T, \tag{5.14b}$$

and for N *odd*

$$G(e^{j\omega T}) = e^{-j\omega T(N-1)/2} \,.\, 2\,\mathrm{Re}\left[\sum_{n=0}^{(N-1)/2} \tilde{a}_n z^{-[n-(N-1)/2]} \right]\Bigg|_{z=e^{j\omega T}} = \tag{5.15a}$$

$$= e^{-j\omega T(N-1)/2}\left[2 \sum_{n=0}^{(N-3)/2} a_n \cos\left(n - \frac{N-1}{2}\right)\omega T + a_{(N-1)/2} \right] \tag{5.15b}$$

where $\tilde{a}_n = a_n$ for $n = 0, 1, \ldots, \dfrac{N-3}{2}$ and $\tilde{a}_{(N-1)/2} = a_{(N-1)/2}/2$.

For numerical computations the relations (5.14a) and (5.15a), in which we apply the Collatz scheme to polynomials of degree $N/2 - 1$ and $(N-1)/2$, respectively, are more advantageous.

In both cases, the expression in the square bracket is a real function of the radian frequency while the term in front of the bracket has linear phase. Consequently, for the phase response we have $\varphi(\omega) = \arg G(e^{j\omega T}) = -\omega T(N-1)/2$, for group delay we have $\tau(\omega) = T(N-1)/2$. This implies that *digital filters with finite and symmetric impulse response have linear phase response.* Similar relations and conclusions can be derived for the antisymmetric impulse response for which we have $a_n = -a_{N-n-1}$.

5.2. Analysis of Digital Filters

In the previous section the realization of transfer functions of digital filters was discussed. Several effective algorithms for the computation of frequency characteristics of digital filters from transfer functions were presented. However, in a number of applications we encounter the inverse problem in which the structure of a digital filter is given, including the numerical constants, and from the structure we have to determine not only the time response but also the transfer function and the frequency characteristics. For instance, we find problems formulated in this manner in the analysis of multiloop structures where a simple correspondence of the coefficients of the transfer function with the coefficients of the structure does not exist, or when simulating digital filters working with limited accuracy of the coefficients as well as of the intermediate results. In this section our aim is therefore to review several possible approaches to the problem and their mutual relations.

5.2.1. *Matrix Representation of Digital Filters*

The *structure of a digital filter* can be described similarly as when representing the performance of a continuously working filter by a system of equations of the node voltages or the loop currents. In contradistinction to the system of integro-differential equations obtained in the case of an analog filter, the representation of a digital filter leads to a system of difference equations. We demonstrate the procedure by the direct realization of transfer function (5.2) in the 1st canonic form. For the transform of its response, relation (5.3b) holds and it can be modified to the form

$$H(z) = a_0\,F(z) + \sum_{i=1}^{s} z^{-i}[a_i\,F(z) - b_i\,H(z)] =$$
$$= a_0F + z^{-1}(a_1F - b_1H + z^{-1}(a_2F - b_2H + + z^{-1} \dots z^{-1}(a_sF + b_sH \dots))), \tag{5.16}$$

where the simpler notation $H = H(z)$ and $F = F(z)$ is used. Denoting the expressions in the individual parentheses as transforms of auxiliary signals – the intermediate results H_i – we obtain the following system of first order equations in the variable z^{-1}:

$$
\begin{aligned}
H_1 &= a_s F - b_s H\,, \\
H_2 &= a_{s-1} F - b_{s-1} H + z^{-1} H_1\,, \\
\vdots \quad & \quad \vdots \\
H_{s-1} &= a_1 F - b_1 H + z^{-1} H_{s-2}\,, \\
H_s &= a_0 F + z^{-1} H_{s-1}\,.
\end{aligned}
\tag{5.17}
$$

System (5.17) can be written in matrix form as

$$
\begin{bmatrix}
-1 & 0\ 0 \ldots 0 & 0 & -b_s \\
z^{-1} & -1\ 0 \ldots 0 & 0 & -b_{s-1} \\
\vdots & & & \\
0 & 0\ 0 \ldots z^{-1} & -1 & -b_1 \\
0 & 0\ 0 \ldots 0 & z^{-1} & -1
\end{bmatrix}
\begin{bmatrix}
H_1 \\ H_2 \\ \vdots \\ H_{s-2} \\ H_{s-1} \\ H_s
\end{bmatrix}
= F
\begin{bmatrix}
-a_s \\ -a_{s-1} \\ \vdots \\ \vdots \\ -a_1 \\ -a_0
\end{bmatrix}
\tag{5.18}
$$

This representation corresponds to the realization of Fig. 53 where the input of every delay block corresponds to the auxiliary signal $\{h_{i,n}\} = \mathscr{Z}^{-1}\{H_i(z)\}$. Upon execution of the inverse $\mathscr{Z}$ transform of relations (5.17) it is possible to apply the obtained recurrent relations directly to the evaluation of the response of the filter to a given input signal $\{f_n\}$, in the same sequence of operations as in the analyzed filter.

The system of equations can be solved applying determinants. With their aid we obtain transfer functions from the input to the outputs of the individual adders

$$
G_i(z) = \frac{H_i}{F} = \frac{D_i}{D} \qquad \text{for} \qquad i = 1, 2, \ldots, s\,,
\tag{5.19a}
$$

where D is the determinant of the system while D_i is the determinant which is obtained from determinant D by replacing the ith column by the right-hand side vector. Since the sth column corresponds to the response $H_s = H$, we have

$$
G_s = \frac{H}{F} = G(z)\,.
\tag{5.19b}
$$

The indicated approach makes it possible to construct a system of equations to a given arbitrary structure which involves delay by integer multiples of the sampling interval T. From this system of equations we can determine transfer functions as ratios of characteristic polynomials, or we can determine zeros and poles of transfer functions with the aid of eigenvalues. Further, the substitution $z^{-1} = \cos \omega T - \mathrm{j} \sin \omega T$ in the matrix representation yields, upon solving repeatedly systems of equations with complex coefficients, the frequency characteristics of the transfer functions from the input to the output of the individual adders.

5.2.2. *Representation of Digital Filters by Signal-Flow Graphs*

In this subsection we resume the considerations of Section 5.1 which led to the representation of digital filters by signal-flow graphs. A signal-flow graph is a useful and clear graphical representation of the relations between given and desired quantities in a digital filter. It can be evaluated or modified with the aid of elementary transformations [26]. To a given structure it is possible to construct a graph, easily and uniquely, exactly as to a given system of linear algebraic and difference first order equations. Conversely, it is uniquely possible to write, to a given graph, a system of equations which are represented by the graph.

However, the principal aim when analyzing or simulating a digital filter is to determine rapidly a sequence of values of the time response, not only at the filter output but also at all points where addition or weighting of signals takes place. For this purpose we need to represent the information on signal flow in the graph with minimal demands on the quantity of data so that the representation be suitable for sequential computation. The notation of the signal flow over the individual functional blocks of a digital filter in *tabular form* proved useful as one possible form of representation.

Let us return to Fig. 50b. To improve the clearness of the graph it is reasonable to introduce several further simplifications in the description of its nodes: for every node $V_1^{(i)}$ the information is sufficient from where the signal is supplied. Since the signals can be either undelayed, i.e. from node $V_1^{(j)}$, or delayed from node $V_2^{(j)}$ $(j = 1, 2, \ldots, m,$ m is the number of blocks), we introduce the convention that the first output of the jth block will be denoted by j while the second – delayed – output by $-j$. The source node is denoted by a zero.

As an example, let us have a digital filter of second order with infinite impulse response, realized in 1st canonic form, whose transfer function is

$$G(z) = \frac{a_0 z^2 + a_1 z + a_2}{z^2 + b_1 z + b_2}. \tag{5.20}$$

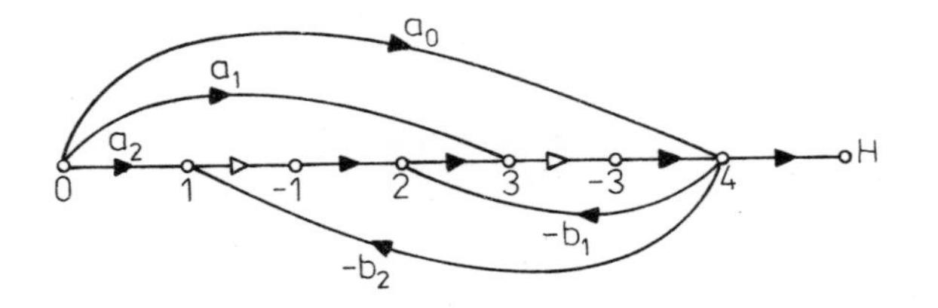

Fig. 60

The corresponding signal-flow graph with the given simplified description is shown in Fig. 60. The graph, and thus the digital filter, can be uniquely represented in tabular form. In consequence of the introduced symbols the table has only digital entries. This is advantageous for computer processing. The solved example corresponding to Fig. 60 is represented by Table 5.1.

TABLE 5.1

Block i	Input $U_1^{(i)}$		Input $U_2^{(i)}$	
	from	Weight $a_1^{(i)}$	from	Weight $a_2^{(i)}$
1	0	a_2	4	$-b_2$
2	-1	1	4	$-b_1$
3	0	a_1	2	1
4	0	a_0	-3	1
Output: 4				

By classification according to various viewpoints, the table yields information concerning the filter, e.g., the number of blocks with and without delay, the number of multipliers, the number of adders, the number and type of loops, etc. Tabular notation is suitable for the representation of classical canonic structures as well as of general structures, namely digital filters with infinite as well as finite impulse response.

Before proceeding with the computation of the time response we present the graphs and tabular representations of several basic structures. The signal-flow graph of Fig. 61 corresponds to the realization of transfer

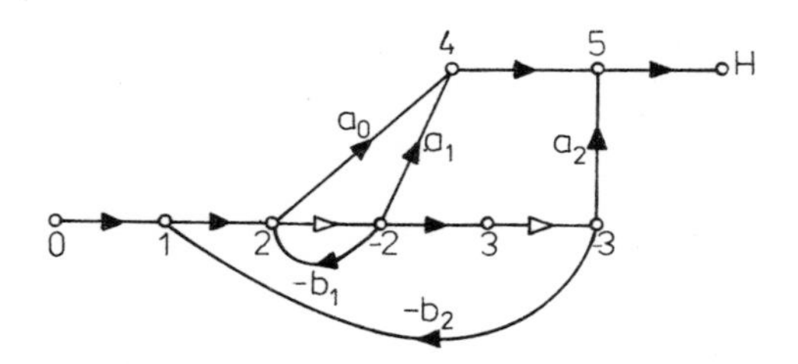

Fig. 61

function (5.20) in the 2nd canonic structure. The corresponding tabular representation is given in Table 5.2. Note that block 3 is degenerated since it has only one input, i.e. its second input branch has zero transmission and the notation of its input node is irrelevant. In these cases we shall write a zero (source node) into the corresponding "from" column.

TABLE 5.2

Block i	Input $U_1^{(i)}$		Input $U_2^{(i)}$	
	from	Weight $a_1^{(i)}$	from	Weight $a_2^{(i)}$
1	0	1	-3	$-b_2$
2	1	1	-2	$-b_1$
3	-2	1	0	0
4	2	a_0	-2	a_1
5	4	1	-3	a_2

Output: 5

Graphs and tables for second order filters with finite impulse response in the 1st and 2nd canonic forms are obtained, in principle, from the above structures where we put $b_1 = b_2 = 0$. In the 1st canonic form this enables to save block 3, but it has no other influence on the character of the graph and of the table. In the case of filters with linear phase which are characterized by the symmetry of the coefficients of the polynomial $P(z)$, i.e. in our case by $a_2 = a_0$, it is possible to achieve still more favourable utilization of the blocks.

Graphs for the 3rd and 4th canonic forms are obtained by the cascade and parallel connection of partial filters in the 1st and 2nd canonic forms.

From a graph, or from a table representing the graph, we can directly write down a system of linear algebraic and difference equations of first order with the aid of which it is possible to determine, to a given input signal $\{f_n\}$, the time response $\{v_{1,n}^{(i)}\}$ at the individual nodes $i = 1, 2, \ldots, m$. However, the basic assumption necessary for the determination of the response is that the corresponding filter be realizable, i.e. that it include no loop without delay. A *filter is realizable* if at least one succession of nodes $\{v_{1,n}^{(i)}\}$ exists which assures that all the necessary node signals be available when proceeding from one node to another. The rule for the determination of the succession of nodes, and thus also for checking the realizability from the table, is the following:

(a) Start from node k which is fed by a signal from node 0 (source node) and by a signal from node $-l$ (delayed), $k = 1, 2, \ldots, m$, $l = 1, 2, \ldots, m$.

(b) Continue with node j which is fed by two signals from the possible nodes $0, k, -k, -l$.

(c) Successively compute the response at nodes fed by: the signal from node 0, signals from nodes $-l$ (delayed), or signals determined in the preceding steps (undelayed).

TABLE 5.3

n	0	1
i	f_0	f_1
4	$v_{1,0}^{(4)} = a_0 f_0$	$v_{1,1}^{(4)} = a_0 f_1 + v_{1,0}^{(3)}$
1	$v_{1,0}^{(1)} = a_2 f_0 - b_2 v_{1,0}^{(4)}$	$v_{1,1}^{(1)} = a_2 f_1 - b_2 v_{1,1}^{(4)}$
2	$v_{1,0}^{(2)} = \quad - b_1 v_{1,0}^{(4)}$	$v_{1,1}^{(2)} = v_{1,0}^{(1)} - b_1 v_{1,1}^{(4)}$
3	$v_{1,0}^{(3)} = a_1 f_0 + v_{1,0}^{(2)}$	$v_{1,1}^{(3)} = a_1 f_1 + v_{1,1}^{(2)}$

For the chosen example described in Table 5.1 the sequence of nodes is $\{4, 1, 2, 3\}$. The computations of the time response at all nodes of the graph under zero initial conditions can be organized in accordance with Table 5.3, derived from Table 5.1. The computation for $n = 2$ is similar as for $n = 1$.

5.2.3. *Transformation of Tabular Representation of a Signal-Flow Graph into a Matrix Equation*

Before proceeding with the subject matter of this subsection let us briefly review the basic principle of the *matrix representation of a digital filter.* A digital filter with one input can be represented by the matrix equation

$$\mathbf{QH} = -F\mathbf{P}, \tag{5.21}$$

where

$$\mathbf{Q} = \begin{bmatrix} a_{11}z^{-1} + b_{11} & a_{12}z^{-1} + b_{12} & \dots & a_{1r}z^{-1} + b_{1r} \\ a_{21}z^{-1} + b_{21} & a_{22}z^{-1} + b_{22} & \dots & a_{2r}z^{-1} + b_{2r} \\ \vdots & & & \\ a_{r1}z^{-1} + b_{r1} & a_{r2}z^{-1} + b_{r2} & \dots & a_{rr}z^{-1} + b_{rr} \end{bmatrix},$$

$${}^{\mathrm{T}}\mathbf{H} = [H_1 \ H_2 \ \dots \ H_r],$$

$${}^{\mathrm{T}}\mathbf{P} = [c_1z^{-1} + d_1 \ \ c_2z^{-1} + d_2 \ \dots \ c_rz^{-1} + d_r].$$

The function $F = F(z)$ is the transform of the input signal. The numbers a_{ik}, b_{ik}, c_k, d_k, $i = 1, 2, \dots, r$, $k = 1, 2, \dots, r$, are coefficients of the digital filter, with the exception of the coefficient b_{kk} whose position corresponds to the node signal H_k and for which we have $b_{kk} = -1$. As a rule, the matrices **Q**, **P** are sparse, i.e. they include many zero elements. The vector ***H*** is constituted by the node signals H_k, including the output signal for $k = r$. By a *node signal* we understand a sum either in the form

$$H_k = \sum_{\substack{v=1 \\ v \neq k}}^{r} (a_{kv}z^{-1} + b_{kv}) H_v + H_k a_{kk} z^{-1} + (c_k z^{-1} + d_k) F \tag{5.22a}$$

or, following the inverse $\mathscr{Z}$ transform in the time domain, in the form

$$h_{k,n} = \sum_{\substack{v=1 \\ v \neq k}}^{r} (a_{kv}h_{v,n-1} + b_{k,v}h_{vn}) + a_{kk}h_{k,n-1} + c_k f_{n-1} + d_k f_n. \tag{5.22b}$$

Obviously, the thus defined matrix notation admits the application of adders with more than two inputs.

The transfer function of the digital filter represented by equation (5.21) from the input to the ith node is given as the ratio of determinants

$$G_i = \frac{H_i}{F} = \frac{\det \mathbf{Q}_i}{\det \mathbf{Q}}, \qquad i = 1, 2, \ldots, r. \tag{5.23}$$

Here det $\mathbf{Q}$ is the determinant of the system while det $\mathbf{Q}_i$ is the determinant which corresponds to the matrix of the system $\mathbf{Q}$ with the ith column replaced by the vector $\mathbf{P}$. Obviously, for $i = r$ we have $G_r = G$. The transfer functions G_i can also be determined by way of the repeated computation of the eigenvalues of the matrices $\mathbf{Q}_i$ and $\mathbf{Q}$.

If we start from the definition of the basic block of a digital filter, we obtain for the transformation of the tabular representation of the signal-flow graph to the matrix representation the following rules:

(a) To every ith block in the graph we assign the ith row of the system of equations (5.21), $i = 1, 2, \ldots, m$.

(b) To every internal node $V_1^{(i)}$ of block i corresponds the ith column of matrix $\mathbf{Q}$.

(c) The transmission of the branch from node 0 (the source node) to node $V_1^{(i)}$ is given by the coefficient d_i of the vector $\mathbf{P}$. For the coefficients c_i we have $c_i \equiv 0$.

In line with these rules the matrix $\mathbf{Q}$ will now have $r = m$ rows and columns and $b_{ii} = -1$ will again hold. The transmission of branch $a_1^{(i)}$, or $a_2^{(i)}$, between node k and block i corresponds to the coefficient b_{ik} in matrix $\mathbf{Q}$. Similarly, the transmission $a_2^{(i)}$, or $a_1^{(i)}$, between node $-k$ and block i corresponds to the coefficient a_{ik} in matrix $\mathbf{Q}$. Since we admit only two inputs per block, the system of equations will as a rule be formulated according to a graph of higher order than if it would be constructed directly to a filter without this constraint. However, for its reduction it is possible to use the procedure of [24]; it will also be applied in the next subsection.

As an example we again use the digital filter of Fig. 60 represented by Table 5.1. The application of the above rules yields the matrix equation (5.24) which represents the realization of transfer function (5.20) in the 1st canonic form:

$$\begin{bmatrix} -1 & 0 & 0 & -b_2 \\ z^{-1} & -1 & 0 & -b_1 \\ 0 & 1 & -1 & 0 \\ 0 & 0 & z^{-1} & -1 \end{bmatrix} \begin{bmatrix} V_1^{(1)} \\ V_1^{(2)} \\ V_1^{(3)} \\ V_1^{(4)} \end{bmatrix} = -F \begin{bmatrix} a_2 \\ 0 \\ a_1 \\ a_0 \end{bmatrix} \tag{5.24}$$

5.2.4. *Relation Between Matrix and State Representations of Digital Filters*

The system of equations representing the signal-flow graph obtained in the preceding subsection expresses the mutual relation of the internal node signals $V_1^{(i)}$, $i = 1, 2, \ldots, m$, and their dependence on the input signal of the source node F. The system of equations in implicit form and matrix notation is

$$\left[\mathbf{d}_i \quad z^{-1}\mathbf{a}_{ik} + \mathbf{b}_{ik}\right] \begin{bmatrix} F \\ \mathbf{V} \end{bmatrix} = \mathbf{0}, \tag{5.25}$$

where $\mathbf{d}_i$ is a column vector, $\mathbf{a}_{ik}$ and $\mathbf{b}_{ik}$ are square matrices, F is a scalar, and $\mathbf{V} = {}^{\mathrm{T}}[V_1, V_2, \ldots, V_m]$ is the column vector of undelayed node signals for which the simplified notation $V_i = V_1^{(i)}$ was introduced. In the matrix $\mathbf{b}_{ik}$ we have $b_{ii} = -1$ for elements on the main diagonal.

The state representation of a digital filter expresses its properties with the aid of *state signals*. They are the quantities preserved in the delay elements, in our case the delayed signals $V_2^{(i)} = z^{-1}V_1^{(i)} = z^{-1}V_i$. Since, as a rule, only a part of the blocks includes delay elements, the number of state signals is $s \leqq m$ and we introduce for them a new symbol W_j, $j = 1, 2, \ldots, s$. They are obtained from the internal signals V_i by the linear transform

$$\mathbf{W} = \begin{bmatrix} W_1 \\ W_2 \\ \vdots \\ W_s \end{bmatrix} = z^{-1}\mathbf{E} \begin{bmatrix} V_1 \\ V_2 \\ \vdots \\ V_m \end{bmatrix} = z^{-1}\mathbf{EV}. \tag{5.26}$$

Here, $\mathbf{E}$ is generally a rectangular transformation matrix of the (s, m) type. The matrix $\mathbf{E}$ includes in every row only one nonzero element which is equal to one, this in column k which corresponds to the kth block in which a delay element is used.

From equations (5.25) and (5.26) we now construct an implicit system of equations to which we add an equation defining the output signal H.

$$\begin{bmatrix} d_0 & -1 & \underset{1,s}{\hat{\mathbf{a}}_k} & \underset{1,m}{\mathbf{b}_k} \\ \underset{s,1}{\mathbf{0}} & \underset{s,1}{\mathbf{0}} & \underset{s,s}{-\mathbf{1}} & \underset{s,m}{z^{-1}\mathbf{E}} \\ \underset{m,1}{\mathbf{d}_i} & \underset{m,1}{\mathbf{0}} & \underset{m,s}{-\hat{\mathbf{a}}_{ik}} & \underset{m,m}{\mathbf{b}_{ik}} \end{bmatrix} \begin{bmatrix} F \\ H \\ \mathbf{W} \\ \mathbf{V} \end{bmatrix} = \mathbf{r} \begin{bmatrix} F \\ H \\ \mathbf{W} \\ \mathbf{V} \end{bmatrix} = \mathbf{0}. \tag{5.27a}$$

Below the symbols of the submatrices we indicate the types of the matrices. The matrices $\underset{1,s}{\hat{\mathbf{a}}_k}$ and $\underset{m,s}{\hat{\mathbf{a}}_{ik}}$ are obtained from the matrices $\underset{1,m}{\mathbf{a}_k}$ and $\underset{m,m}{\mathbf{a}_{ik}}$, respectively, in which all zero columns are omitted; $\mathbf{0}$ is the zero matrix, $\mathbf{1}$ is the unit matrix.

To obtain state equations from the augmented system of signal-flow equations the matrix $\mathbf{r}$ is divided into submatrices as follows

$$\left[\begin{array}{c|c} \mathbf{r}_{11} & \mathbf{r}_{12} \\ \hline \mathbf{r}_{21} & \mathbf{r}_{22} \end{array}\right] \left[\begin{array}{c} F \\ H \\ \mathbf{W} \\ \hline \mathbf{V} \end{array}\right] = \mathbf{0}, \tag{5.27b}$$

and the vector of internal nonstate signals $\mathbf{V}$ is excluded. Under the assumption that an inverse matrix exists to $\mathbf{r}_{22}$, we obtain the following implicit matrix system which includes only state variables

$$[\mathbf{r}_{11} - \mathbf{r}_{12}\mathbf{r}_{22}^{-1}\mathbf{r}_{21}] \begin{bmatrix} F \\ H \\ \mathbf{W} \end{bmatrix} = \mathbf{s} \begin{bmatrix} F \\ H \\ \mathbf{W} \end{bmatrix} = \mathbf{0}. \tag{5.28}$$

Further, we write out equation (5.28) into two matrix equations for the output signal H and the vector of state variables

$$s_{11}F + s_{12}H + \mathbf{s}_{13}\mathbf{W} = 0, \tag{5.29a}$$

$$\mathbf{s}_{21}F + \mathbf{s}_{22}H + \mathbf{s}_{23}\mathbf{W} = \mathbf{0}. \tag{5.29b}$$

Equation (5.29a) yields the *second state equation* for the output signal

$$H = -\frac{\mathbf{s}_{13}}{s_{12}}\mathbf{W} - \frac{s_{11}}{s_{12}}F. \tag{5.30a}$$

Substituting into (5.29b) and modifying we obtain the *first state equation*

$$z\mathbf{W} = z\left(\mathbf{s}_{23} + \mathbf{1} - \frac{\mathbf{s}_{22}\mathbf{s}_{13}}{s_{12}}\right)\mathbf{W} + z\left(\mathbf{s}_{21} - \frac{\mathbf{s}_{22}s_{11}}{s_{12}}\right)F. \tag{5.30b}$$

Comparison with the state equations in compact form

$$\begin{aligned} z\mathbf{W} &= \mathbf{A}\mathbf{W} + \mathbf{B}F, \\ H &= \mathbf{C}\mathbf{W} + DF, \end{aligned} \tag{5.31}$$

yields

$$\mathbf{A} = z\left(\mathbf{s}_{23} + \mathbf{1} - \frac{\mathbf{s}_{22}\mathbf{s}_{13}}{s_{12}}\right), \tag{5.32a}$$

$$\mathbf{B} = z\left(\mathbf{s}_{21} - \frac{\mathbf{s}_{22}s_{11}}{s_{12}}\right), \tag{5.32b}$$

$$\mathbf{C} = -\frac{\mathbf{s}_{13}}{s_{12}}, \tag{5.32c}$$

$$D = -\frac{s_{11}}{s_{12}}. \tag{5.32d}$$

The inverse $\mathscr{Z}$ transform leads to the state difference equations from which it is possible to compute successively, from the initial states $\mathbf{w}_0 = \mathscr{Z}^{-1}\{\mathbf{W}\}|_{n=0}$, the state signals and the output signal.

From the state equations (5.32) we obtain, by the elimination of the state vector, the transfer function

$$G = \frac{H}{F} = \mathbf{C}[z\mathbf{1} - \mathbf{A}]^{-1}\mathbf{B} - D. \tag{5.33}$$

For illustration, let us construct the state equations for the digital filter of Fig. 60 for which we have obtained matrix equation (5.24) in the previous subsection. This equation rewritten into implicit form is

$$\begin{bmatrix} a_2 & -1 & 0 & 0 & -b_2 \\ 0 & z^{-1} & -1 & 0 & -b_1 \\ a_1 & 0 & 1 & -1 & 0 \\ a_0 & 0 & 0 & z^{-1} & -1 \end{bmatrix} \begin{vmatrix} F \\ V_1 \\ V_2 \\ V_3 \\ V_4 \end{vmatrix} = \mathbf{0}.$$

The number of delay elements is $s = 2$, namely in blocks $i = 1$ and $i = 3$. Consequently, the state signals are obtained from the internal signals V_i, $i = 1, 2, 3, 4$, by the linear transformation

$$\mathbf{W} = \begin{bmatrix} W_1 \\ W_2 \end{bmatrix} = z^{-1}\mathbf{E}\mathbf{V} = \begin{bmatrix} z^{-1} & 0 & 0 & 0 \\ 0 & 0 & z^{-1} & 0 \end{bmatrix} \begin{bmatrix} V_1 \\ V_2 \\ V_3 \\ V_4 \end{bmatrix}.$$

The system of equations (5.27a) which contains the sink node, internal and also state variables, will for our example acquire the form

$$\begin{bmatrix} 0 & -1 & 0 & 0 & 0 & 0 & 0 & 1 \\ 0 & 0 & -1 & 0 & z^{-1} & 0 & 0 & 0 \\ 0 & 0 & 0 & -1 & 0 & 0 & z^{-1} & 0 \\ a_2 & 0 & 0 & 0 & -1 & 0 & 0 & -b_2 \\ 0 & 0 & 1 & 0 & 0 & -1 & 0 & -b_1 \\ a_1 & 0 & 0 & 0 & 0 & 1 & -1 & 0 \\ a_0 & 0 & 0 & 1 & 0 & 0 & 0 & -1 \end{bmatrix} \begin{bmatrix} F \\ H \\ W_1 \\ W_2 \\ V_1 \\ V_2 \\ V_3 \\ V_4 \end{bmatrix} = \mathbf{0}.$$

By (5.27b) the brief notation is

$$\begin{bmatrix} \mathbf{r}_{11} & \mathbf{r}_{12} \\ \mathbf{r}_{21} & \mathbf{r}_{22} \end{bmatrix} \begin{bmatrix} F \\ H \\ \mathbf{W} \\ \mathbf{V} \end{bmatrix} = \mathbf{r} \begin{bmatrix} F \\ H \\ \mathbf{W} \\ \mathbf{V} \end{bmatrix} = \mathbf{0}.$$

If we exclude the vector $\mathbf{V}$ of internal signals, we obtain for

$$\mathbf{r}_{22}^{-1} = \begin{bmatrix} -1 & 0 & 0 & b_2 \\ 0 & -1 & 0 & b_1 \\ 0 & -1 & -1 & b_1 \\ 0 & 0 & 0 & -1 \end{bmatrix}$$

the system (5.29) which contains only state signals:

$$\left[\begin{array}{c|cc|c} a_0 & -1 & 0 & 1 \\ \hline a_2 z^{-1} - a_0 b_2 z^{-1} & 0 & -1 & -b_2 z^{-1} \\ a_1 z^{-1} - a_0 b_1 z^{-1} & 0 & z^{-1} & -1 - b_1 z^{-1} \end{array}\right] \begin{bmatrix} F \\ \hline H \\ \mathbf{W} \end{bmatrix} =$$

$$= \begin{bmatrix} s_{11} & s_{12} & \mathbf{s}_{13} \\ \mathbf{s}_{21} & \mathbf{s}_{22} & \mathbf{s}_{23} \end{bmatrix} \begin{bmatrix} F \\ H \\ \mathbf{W} \end{bmatrix} = \mathbf{0}.$$

Applying relations (5.32) we then obtain the state equations

$$z\mathbf{W} = \begin{bmatrix} 0 & -b_2 \\ 1 & -b_1 \end{bmatrix} \mathbf{W} + \begin{bmatrix} a_2 - a_0 b_2 \\ a_1 - a_0 b_1 \end{bmatrix} F,$$

$$H = \begin{bmatrix} 0 & 1 \end{bmatrix} \mathbf{W} + a_0 F,$$

from which the initial transfer function is easily constructed by (5.33).

5.2.5. *Computation of the Transfer Function of a Digital Filter from Its Time Response*

Till now we have dealt with the representation of digital filters by signal-flow graphs with the aim of determining the time response to a given input signal at the individual nodes. However, when simulating digital filters we have to solve, for instance, problems of stability, problems of sensitivity of the zeros and poles to the quantization of the coefficients, the effect of the quantization of the coefficients on the frequency characteristics. Frequently this cannot be solved without determining the transfer function of the filter. For the majority of the classical canonic structures (with the exception of the 4th canonic structure) the coefficients of the transfer function are as a rule included directly in the structure; however, for the other structures the coefficients of the transfer function are functions of the multiplicative coefficients in the branches of the filter. In these cases we have to identify the transfer function from the structure. The classical approach using determinants or eigenvalues was indicated in Subsection 5.2.1; however, this approach assumes the construction of a matrix equation to a given signal-flow graph. This procedure, current in the analysis of continuous systems, can be replaced when analyzing digital filters by the identification of the transfer function from its exact impulse response. This is easily determined – in contradistinction to continuous systems – by recursive computation directly from the signal-flow graph. Below a procedure is outlined for the *identification of the transfer function* from the impulse response which is more suitable for our applications than the general approach.

Let the transfer function

$$G(z) = \frac{a_0 z^s + a_1 z^{s-1} + \ldots + a_s}{z^s + b_1 z^{s-1} + \ldots + b_s} = \frac{P(z)}{Q(z)} = \mathscr{Z}\{g_n\} \tag{5.34}$$

be given. All its equivalent realizations have the same response $\{h_n\}$ for a given input signal $\{f_n\}$, thus they have the same response as its direct realization in the 1st canonic form. For the impulse response of the corresponding digital filter we have

$$g_n = \mathscr{Z}^{-1}\{G(z)\} = a_n - \sum_{i=1}^{n} b_i g_{n-i}, \tag{5.35}$$

where we put $a_n = b_n = 0$ for $n > s$. This relation can be written in matrix form (see Subsection 4.1.5):

$$\left[\begin{array}{cccc|ccc} 1 & 0 & \ldots & 0 & 0 & \ldots & 0 \\ 0 & 1 & \ldots & 0 & -g_0 & \ldots & 0 \\ \vdots & \vdots & & \vdots & \vdots & & \vdots \\ 0 & 0 & \ldots & 1 & -g_{s-1} & \ldots & -g_0 \\ \hline 0 & 0 & \ldots & 0 & -g_s & \ldots & -g_1 \\ \vdots & \vdots & & \vdots & \vdots & & \vdots \\ 0 & 0 & \ldots & 0 & -g_{2s} & \ldots & -g_{s+1} \end{array}\right] \left[\begin{array}{c} a_0 \\ a_1 \\ \vdots \\ a_s \\ \hline b_1 \\ \vdots \\ b_s \end{array}\right] = \left[\begin{array}{c} g_0 \\ g_1 \\ \vdots \\ g_s \\ \hline g_{s+1} \\ \vdots \\ g_{2s+1} \end{array}\right]. \tag{5.36}$$

With respect to the form of the matrix equation and to the fact that some types of digital filters are realizations of transfer functions which have certain coefficients identically equal to zero, it is advantageous not to determine the coefficients of $P(z)$ and $Q(z)$ simultaneously but to determine the coefficients of the denominator $Q(z)$ in the first step and the coefficients of the numerator from the initial values of the impulse response in the second step. This two-step identification procedure is advantageous when computing the response from the signal-flow graph since the computation yields the response at all nodes simultaneously. Thus, we have the possibility to obtain, by one solution of the system of s equations in s unknowns, the transfer functions G_i from the source node to all nodes $i = 1, 2, \ldots, m$ (see relation (5.23)). Under the assumption of the existence of the inverse matrix all the coefficients of the polynomial $Q(z)$ are obtained by the solution of the partial system of equations

$$\begin{bmatrix} b_1 \\ b_2 \\ \vdots \\ b_s \end{bmatrix} = - \begin{bmatrix} g_s & g_{s-1} & \cdots & g_1 \\ \vdots & \vdots & & \vdots \\ g_{2s} & g_{2s-1} & \cdots & g_{s+1} \end{bmatrix}^{-1} \begin{bmatrix} g_{s+1} \\ \vdots \\ g_{2s+1} \end{bmatrix} . \tag{5.37}$$

If we express $P(z)$ from equation (5.34), we obtain

$$P(z) = Q(z) \sum_{n=0}^{+\infty} g_n z^{-n} , \tag{5.38}$$

which yields for the sequence of coefficients of the polynomial $P(z)$

$$\begin{aligned} a_0 &= g_0 , \\ a_1 &= g_1 + b_1 g_0 , \\ a_2 &= g_2 + b_1 g_1 + b_2 g_0 , \\ \vdots \quad & \quad \vdots \\ a_s &= g_s + b_1 g_{s-1} + \ldots + b_s g_0 . \end{aligned} \tag{5.39}$$

If we put $P(z) = 0$ and $Q(z) = 0$, we obtain zeros and poles of the transfer function of a digital filter. For digital filters (designed first of all with the aid of the bilinear transformation) multiple zeros, or poles at the point $z = \pm 1$, occur frequently. These determine the properties of the filter at zero or half the sampling frequency. Therefore, it is reasonable to verify the existence and multiplicity of these roots before solving the equation. This is achieved by the substitution of the variable $z = z' + 1$, or $z = z' - 1$, with the aid of the Horner scheme, after which the number of zero coefficients starting from the absolute term determines the multiplicity of the root at the point $z = 1$, or $z = -1$, respectively. In case that such roots are present we remove the corresponding root factors and proceed with the solution of the equation, of lower degree now, with roots $z_\nu \neq \pm 1$. The solution of a number of examples showed that this approach to the determination of the zeros and poles by the solution of polynomial equations is at least as suitable as the direct computation of the eigenvalues by matrix methods.

5.2.6. *Computation of Frequency Characteristics of Digital Filters from Their Time Response*

Similarly as when measuring the amplitude frequency response of a digital filter by evaluating the ratio of the amplitude of its steady-state response to the amplitude of its harmonic input signal for the chosen frequency, we can also determine the individual points of the frequency response when analyzing a digital filter from samples of signals on its input and output. The procedure is the following:

For the response of a stable digital filter, with transfer function $G(z)$ at steady state, to the complex harmonic signal

$$f_n = \mathrm{e}^{\mathrm{j}(\omega Tn + \beta)} = \cos(\omega Tn + \beta) + \mathrm{j}\sin(\omega Tn + \beta) \tag{5.40}$$

we have, for $n \geqq n_0$,

$$\begin{aligned} h_n &= \mathrm{e}^{\mathrm{j}(\omega Tn + \beta)}\, G(\mathrm{e}^{\mathrm{j}\omega T}) = A(\omega)\, \mathrm{e}^{\mathrm{j}[\omega Tn + \beta + \varphi(\omega)]} = \\ &= A(\omega)\left\{\cos\left[\omega Tn + \beta + \varphi(\omega)\right] + \mathrm{j}\sin\left[\omega Tn + \beta + \varphi(\omega)\right]\right\}, \end{aligned} \tag{5.41}$$

where n_0 is a suitably chosen number under which it is possible to consider the response to be stationary. The value of the amplitude response

$A(\omega) = A$ and the value of the phase response $\varphi(\omega) = \varphi$ are determined by the solution of two equations obtained from the response to the real signal $\cos(\omega Tn + \beta)$ for given radian frequency ω for n and $(n+1)$. We introduce the notation

$$u_n = \cos(\omega Tn + \beta), \qquad v_n = \sin(\omega Tn + \beta).$$

For the response to the signal $\{u_n\}$ we have

$$\begin{aligned} h_n &= A(u_n \cos\varphi - v_n \sin\varphi), \\ h_{n+1} &= A(u_{n+1} \cos\varphi - v_{n+1} \sin\varphi), \end{aligned} \tag{5.42}$$

which yields

$$\operatorname{Re} G = A\cos\varphi = \frac{h_n v_{n+1} - h_{n+1} v_n}{u_n v_{n+1} - u_{n+1} v_n} \tag{5.43a}$$

for the real component of the transfer function and

$$\operatorname{Im} G = A\sin\varphi = \frac{h_n u_{n+1} - h_{n+1} u_n}{u_n v_{n+1} - u_{n+1} v_n} \tag{5.43b}$$

for the imaginary component. The common denominator yields, upon modification, $u_n v_{n+1} - u_{n+1} v_n = \sin(\omega T + \beta) = v_1$. Then the desired values of the frequency characteristics are

$$A = A(\omega) = \sqrt{[(\operatorname{Re} G)^2 + (\operatorname{Im} G)^2]}, \tag{5.44a}$$

$$\varphi = \varphi(\omega) = 2\arctan\frac{\operatorname{Im} G}{A + \operatorname{Re} G}, \qquad \varphi \in [0, \pi]. \tag{5.44b}$$

The value of the transfer function for $\omega = 0$, zero frequency transmission, is equal to the value of the steady-state process for $\{f_n\} = 1$, i.e. for the input signal in the form of the unit step. For the stable system we then have

$$\lim_{n \to +\infty} h_n = \lim_{z \to 1+} (z-1)\frac{z}{z-1} G(z) = G(1), \tag{5.45}$$

and thus

$$A = A(0) = |G(1)| = |\lim_{n \to +\infty} h_n|, \tag{5.46a}$$

$$\varphi = \varphi(0) = \arg G(1) = 0 \quad \text{for} \quad \lim_{n \to +\infty} h_n > 0,$$
$$= \pi \quad \text{for} \quad \lim_{n \to +\infty} h_n < 0. \tag{5.46b}$$

The characteristic of group delay is determined from the definitorical relation $\tau = -\mathrm{d}\varphi/\mathrm{d}\omega$ in which we approximate the derivative by the difference. A compromise between simplicity and accuracy leads to the application of the Stirling formula (see Subsection 4.3.2).

The assumption for computing the frequency characteristics in this manner is the stability of the response $\{h_n\}$. The verification of the steady-state of the response, which may be of oscillatory character, is replaced by the verification of the steady-state of the sequence $\{A_n\}$, determined from the sequence $\{h_n\}$, for which $\lim_{n \to +\infty} A_n = A$ holds. The computation of $\{h_n\}$ is concluded for such $n = \nu N$, $\nu = 1, 2, \ldots$, for which either the inequality $|A_{\nu N} - A_{\nu N+1}| \leqq \varepsilon$ for $A_{\nu N} > \varepsilon$ or the inequality $A_{\nu N} \leqq \varepsilon$ is satisfied. The number N is a suitably chosen number (e.g., 20), while ε is the chosen error of the module (e.g., 10^{-3}). The computation of the response $\{h_n\}$ is performed in line with Subsection 5.2.2. Here we pass from one frequency to the other without clearing the internal signals. This decreases the time of reaching the steady-state of the responses for the individual frequencies. Since we compute simultaneously the behaviour $\{v_{1,n}^{(i)}\}$ at all nodes of the graph it is obvious that it is possible, almost simultaneously, to determine the frequency characteristics from the input to all the nodes of the graph. The advantage of this approach to the computation of the frequency characteristics consists precisely in this fact. The comparison of computer time necessary for the evaluation of the frequency characteristics shows that the described procedure has approximately the same demands, for an average digital filter, as the approach consisting in the determination of zeros and poles.

Chapter 6

Application of the $\mathscr{Z}$ Transform to the Analysis of Discrete Signals

6.1. $\mathscr{Z}$ Transform and the Discrete Fourier Transform

By the analysis of signals we usually understand the investigation of their spectral properties. If we start from the conception that every discrete signal can be interpreted as the response of a specific system to the unit impulse, then the methods of the analysis of digital filters can be applied also to the analysis of discrete signals. However, it is also possible to proceed the other way round: the existing methods of the analysis of signals may be applied to the response of a digital filter and its properties can be investigated without the knowledge of its transfer function, i.e. as a matter of fact, it can be identified in the frequency domain. The discrete Fourier transform [10], [29], [32] is the most important method of the analysis of signals. It became an effective tool for the solution of numerous problems thanks to the algorithm of the fast Fourier transform. Now, we attempt to summarize the individual approaches to the analysis of signals and their connections in relation to the theory and application of the $\mathscr{Z}$ transform.

The *discrete Fourier transform* (DFT), or more exactly the finite DFT, is defined for a *periodic sequence* $\{\tilde{f}_n\}$, i.e. for a sequence for which we have $\tilde{f}_n = \tilde{f}_{n+iN}$, $i = 0, \pm 1, \pm 2, \ldots$, by the formula

$$\tilde{F}_k = \sum_{n=0}^{N-1} \tilde{f}_n z^{-\mathrm{j}nk(2\pi/N)}, \qquad k = 0, 1, 2, \ldots, (N-1). \tag{6.1}$$

For the inverse discrete Fourier transform we then have

$$\tilde{f}_n = \frac{1}{N} \sum_{k=0}^{N-1} \tilde{F}_k \mathrm{e}^{\mathrm{j}kn(2\pi/N)}, \qquad n = 0, 1, 2, \ldots, (N-1). \tag{6.2}$$

The sequence $\{\tilde{F}_k\}$ is a complex sequence of values of the discrete Fourier transform and it is also *periodic*, i.e. we have $\tilde{F}_k = \tilde{F}_{k+iN}$, $i = 0, \pm 1, \pm 2, \ldots$. In the sequel, for relations (6.1) and (6.2) we shall use the symbolic notation

$$\tilde{F}_k = \mathscr{D}_N\{\tilde{f}_n\},$$
$$\tilde{f}_n = \mathscr{D}_N^{-1}\{\tilde{F}_k\}.$$

From the periodic sequence $\{\tilde{f}_n\}$ we form a finite sequence by considering its one period, i.e. we put

$$\begin{aligned} f_n &= \tilde{f}_n \quad &&\text{for} \quad n = 0, 1, \ldots, (N-1), \\ &= 0 \quad &&\text{for} \quad \text{the other } n. \end{aligned}$$

Then the $\mathscr{Z}$ transform of this sequence is

$$F(z) = \sum_{n=0}^{N-1} f_n z^{-n}, \tag{6.3}$$

whence, by comparison with (6.1), it follows that

$$\tilde{F}_k = F(z)\Big|_{z = e^{j(2\pi/N)k} = z_k} = F(e^{j(2\pi/N)k}), \quad k = 0, 1, \ldots, (N-1). \tag{6.4}$$

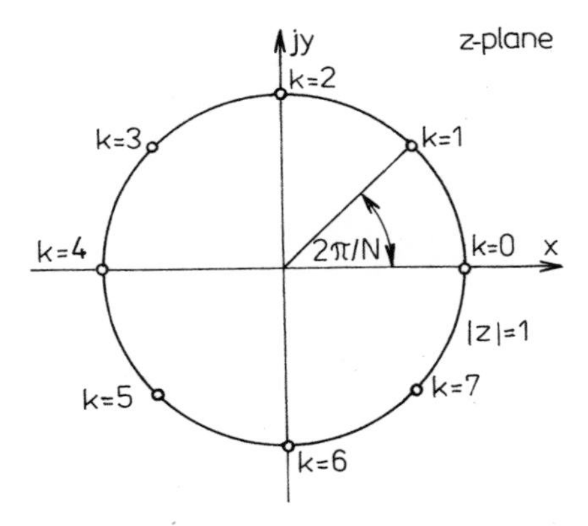

Fig. 62

We see that the discrete Fourier transform of a finite sequence $\{f_n\}$ is given by the complex sequence of samples of the $\mathscr{Z}$ transform $F(z)$ for $z = e^{j\omega T}$ – thus the Fourier transform of a sequence (see Subsection 3.2.2) – equidistantly sampled around the unit circle, i.e. for $\omega_k = (2\pi/N)\,k$, $k = 0, 1, \ldots, (N-1)$. In Fig. 62, sampling along the unit circle is shown for $N = 8$.

Now, let us represent the transform $F(z)$ with the aid of the discrete Fourier transform $\{\tilde{F}_k\}$. If we substitute into (6.3) for $\tilde{f}_n$ by (6.2), modification leads to

$$F(z) = \sum_{n=0}^{N-1} \frac{1}{N} \sum_{k=0}^{N-1} \tilde{F}_k e^{jkn(2\pi/N)} z^{-n} =$$

$$= \sum_{k=0}^{N-1} \tilde{F}_k \frac{1}{N} \sum_{n=0}^{N-1} e^{jkn(2\pi/N)} z^{-n} =$$

$$= \frac{z^N - 1}{N z^N} \sum_{k=0}^{N-1} \tilde{F}_k \frac{z}{z - e^{j(2\pi/N)k}} . \tag{6.5}$$

Thus we have arrived at a relation which is identical with relation (5.11) for the realization of the transfer function of a filter with finite impulse response on the principle of frequency sampling. For points on the unit circle, i.e. for $z = e^{j\omega T}$, trigonometric modifications yield the Fourier transform of the finite sequence, i.e.

$$F(e^{j\omega T}) = \frac{e^{-j\omega T(N-1)/2}}{N} \sum_{k=0}^{N-1} \tilde{F}_k e^{-j(\pi/N)k} \frac{\sin \dfrac{\omega T N}{2}}{\sin\left(\dfrac{\omega T}{2} - \dfrac{\pi}{N} k\right)} . \tag{6.6}$$

It is obvious that the continuous function $F(e^{j\omega T})$ can be constructed from the sequence $\{\tilde{F}_k\}$ with the help of the frequency interpolation function

$$S(\omega, k) = \frac{e^{-j\omega T(N-1)/2}}{N} e^{-j(\pi/N)k} \frac{\sin \dfrac{\omega T N}{2}}{\sin\left(\dfrac{\omega T}{2} - \dfrac{\pi}{N} k\right)} . \tag{6.7}$$

Now, we concentrate our attention to the time domain and show the consequence of the application of DFT to the infinite aperiodic sequence $\{f_n\}$. For this purpose we substitute for $\{\tilde{F}_k\}$ in relation (6.2) the values of the $\mathscr{Z}$ transform obtained by its equidistant sampling around the unit circle in accordance with (6.4). For a periodic sequence we then have

$$\tilde{f}_n = \frac{1}{N} \sum_{k=0}^{N-1} \tilde{F}_k e^{jkn(2\pi/N)} = \frac{1}{N} \sum_{k=0}^{N-1} \sum_{m=0}^{+\infty} f_m e^{-j(2\pi/N)km} e^{j(2\pi/N)kn} =$$

$$= \frac{1}{N} \sum_{k=0}^{N-1} \sum_{m=0}^{+\infty} f_m e^{j(2\pi/N)k(n-m)} . \tag{6.8}$$

Interchanging the order of summation and applying relation

$$\sum_{k=0}^{N-1} e^{j(2\pi/N)k(n-m)} = N \qquad \text{for} \quad n - m = iN\,,$$
$$= 0 \qquad \text{for} \quad n - m \neq iN\,, \quad i = 0, \pm 1, \pm 2, \ldots,$$

we obtain

$$\tilde{f}_n = \sum_{i=-\infty}^{+\infty} f_{n+iN}\,. \tag{6.9}$$

Consequently, if the sequence $\{f_n\}$ is not finite and $f_n = 0$ does not hold for $n \geqq N$, sampling in the frequency domain gives rise to *aliasing in the time domain.* The phenomenon is an analogue of aliasing in the frequency domain caused by sampling in the time domain as discussed in detail in Subsection 4.2.1.

6.1.1. *Spectral Characteristics of Discrete Signals*

In Subsection 3.2.2, the relation between the $\mathscr{Z}$ transform and the Fourier transform of a sequence was derived. Under certain assumptions which will not be repeated here we obtain the Fourier transform of a sequence by the substitution $z = e^{j\omega T}$ in its $\mathscr{Z}$ transform. *The Fourier transform of a sequence is called the spectrum of the sequence.* The relation of the $\mathscr{Z}$ transform and the discrete Fourier transform was discussed above in this chapter. From the discussion we know under what assumptions and with what consequences is it possible to use the discrete Fourier transform for the computation of the $\mathscr{Z}$ transform of a sequence and thus, following the substitution $z = e^{j\omega T}$, for the computation of the Fourier transform, i.e. of the spectrum of the sequence. Thanks to the algorithm of the fast Fourier transform (FFT) the discrete Fourier transform came to be an effective method of the analysis of signals and thus also of the analysis of digital filters. The application of the fast Fourier transform to the computation of the spectra of signals will not be treated here; we refer to specialized literature [9], [10]. However, we will review some of the notions and procedures which will be applied in the following subsections.

Let us assume that we have determined the spectrum of a discrete signal $\{f_n\}$, thus the Fourier transform which is a complex function of the real variable ω:

$$F(e^{j\omega T}) = F(z)\Big|_{z=e^{j\omega T}} = \operatorname{Re} F(e^{j\omega T}) + j \operatorname{Im} F(e^{j\omega T}) =$$
$$= A(\omega)\, e^{j\varphi(\omega)}\,. \tag{6.10}$$

The function $A(\omega) = |F(e^{j\omega T})|$ is the *amplitude spectrum* while the function $\varphi(\omega) = \arg F(e^{j\omega T})$ is the *phase spectrum of the sequence* $\{f_n\}$. The amplitude spectrum is an *even* function while the phase spectrum is an *odd* function of the radian frequency ω. The function $A^2(\omega) = |F(e^{j\omega T})|^2$ is the *power spectral density*, or briefly the *power spectrum of the sequence* $\{f_n\}$. Frequently we encounter also the notion of *logarithmic spectrum* which is defined by the complex logarithm

$$\hat{F}(e^{j\omega T}) = \ln F(z)\Big|_{z=e^{j\omega T}} = \ln F(e^{j\omega T}) =$$
$$= \ln |F(e^{j\omega T})| + j \arg F(e^{j\omega T}) = \ln A(\omega) + j\,\varphi(\omega)\,. \tag{6.11}$$

The real part of the logarithmic spectrum, i.e. the function $\ln A(\omega)$, is called the *logarithmic amplitude spectrum of the sequence* $\{f_n\}$, the imaginary part of the logarithmic spectrum is identical with the phase spectrum.

The phase spectrum is determined, as a rule, with the aid of the arctan function which yields the principal value of the function $\varphi(\omega)$, with discontinuities at points where the function $\varphi(\omega)$ assumes the values $\pm\pi$. However, in a number of applications we need a continuous function $\varphi(\omega)$. One of the possible ways of *unwrapping the phase spectrum* consists in the integration of the derivative of the phase spectrum $\varphi'(\omega) = d\,\varphi(\omega)/d\omega$,

$$\varphi(\omega) = \int_0^{\omega} \varphi'(u)\, du\,, \tag{6.12}$$

with the initial condition $\varphi(0) = 0$ which follows from the requirement that the function $\varphi(\omega)$ be odd. Integration is performed numerically, with arbitrary accuracy.

The derivative of the phase spectrum is obtained from relation (6.11) as follows:

$$\varphi'(\omega) = \operatorname{Im} \frac{\mathrm{d} \ln F(\mathrm{e}^{\mathrm{j}\omega T})}{\mathrm{d}\omega} = \operatorname{Im} \frac{1}{F(\mathrm{e}^{\mathrm{j}\omega T})} \frac{\mathrm{d} F(\mathrm{e}^{\mathrm{j}\omega T})}{\mathrm{d}\omega} =$$

$$= \frac{\operatorname{Re} F(\mathrm{e}^{\mathrm{j}\omega T}) \operatorname{Im} F'(\mathrm{e}^{\mathrm{j}\omega T}) - \operatorname{Im} F(\mathrm{e}^{\mathrm{j}\omega T}) \operatorname{Re} F'(\mathrm{e}^{\mathrm{j}\omega T})}{|F(\mathrm{e}^{\mathrm{j}\omega T})|^2}, \quad (6.13)$$

where the primes denote differentiation with respect to ω. The differentiation of the spectrum $F(\mathrm{e}^{\mathrm{j}\omega T})$ may be approximately performed numerically (e.g., by the Stirling formula).

A further procedure for the computation of the phase spectrum is derived from the complex logarithmic derivative in the variable z. With the aid of Theorem 12 of Subsection 2.1.8 we have

$$\frac{\mathrm{d}}{\mathrm{d}z} \ln F(z) = \frac{1}{F(z)} \frac{\mathrm{d} F(z)}{\mathrm{d}z} = -\frac{1}{F(z)\, z} \mathscr{Z}\{nf_n\}, \quad (6.14)$$

where $f_n = \mathscr{Z}^{-1}\{F(z)\}$ is the sequence whose spectrum we are looking for. For $\varphi'(\omega)$ it is then possible to write, on the one hand,

$$\varphi'(\omega) = \operatorname{Im} \frac{\mathrm{d}}{\mathrm{d}z} \ln F(z) \frac{\mathrm{d}z}{\mathrm{d}\omega}\bigg|_{z=\mathrm{e}^{\mathrm{j}\omega T}} = \operatorname{Im} \mathrm{j}\, Tz \frac{\mathrm{d}}{\mathrm{d}z} \ln F(z)\bigg|_{z=\mathrm{e}^{\mathrm{j}\omega T}} =$$

$$= T \operatorname{Re}\left[\frac{z}{F(z)} \frac{\mathrm{d} F(z)}{\mathrm{d}z}\right]\bigg|_{z=\mathrm{e}^{\mathrm{j}\omega T}} \quad (6.15\mathrm{a})$$

and, on the other hand,

$$\varphi'(\omega) = \operatorname{Im} \mathrm{j}\, T \frac{-\mathscr{Z}\{nf_n\}}{F(z)}\bigg|_{z=\mathrm{e}^{\mathrm{j}\omega T}} = -T \operatorname{Re} \frac{\mathscr{Z}\{nf_n\}}{\mathscr{Z}\{f_n\}}\bigg|_{z=\mathrm{e}^{\mathrm{j}\omega T}}. \quad (6.15\mathrm{b})$$

Both relations can be evaluated at any arbitrary point by the repeated application of the Collatz scheme according to Subsection 5.1.3; the second relation is suitable for the application of the discrete Fourier transform realized by the FFT algorithm.

In this subsection, fundamental notions of the spectral analysis of signals were introduced and some procedures for the computation of spectral characteristics were presented. The manner of their computation must be adapted both to the physical properties of the signal and to the required number of points and thus to the number of arithmetic operations. If several values of the spectrum are sufficient, e.g. at not equidistant points, we apply the Collatz scheme in accordance with Subsection 5.1.3. This scheme is related to the Goertzel algorithm which can be realized by digital filtering and is discussed in the next subsection. If a larger number of values of the spectrum is required, we use the FFT algorithm. In this case, the number of points N has to be chosen so that $N \geqq M$, where M is the total number of signal samples. A further criterion for the choice of the number of points follows from the required number and position of the frequency points equidistantly distributed around the unit circle with frequency spacing $\Delta f = f_s/N$. Particular cautiousness, and thus particular approaches, is required when computing the spectrum of a nonperiodic infinite signal. In this case we have to process a finite segment of the signal using a suitable window. To obtain a likelihood estimate of the spectrum we frequently average spectra corresponding to successive (or even overlapping) sections of the signal. These methods are discussed in specialized literature [8], [9].

6.1.2. *Time Varying Spectrum and the Goertzel Algorithm*

Till now we have discussed the computation of the spectrum of a given section of a sequence and we were not interested in the variability of the spectrum in time. However, in some applications we are interested in only some points of the spectrum, namely in dependence on the discrete time variable n. Let us formulate this problem with the aid of sequences and of the $\mathscr{Z}$ transform [32], [56].

Let us have a sequence $\{f_n\}$, $n = 0, 1, 2, \ldots$. Let us generate a time sequence of the value of the spectrum — a *sliding value of the spectrum*, i.e. a sequence $\{s_n(z_k)\}$ determined at the point $z_k = e^{j(2\pi/N)k}$, $k = 0, 1, 2, \ldots, (N-1)$, by N terms of the sequence $\{f_n\}$ according to the formula

$$s_n(z_k) = f_n + f_{n-1} z_k^{-1} + \ldots + f_{n-N-1} z_k^{-(N-1)} = \\ = \sum_{m=0}^{N-1} f_{n-m} z_k^{-m} . \tag{6.16a}$$

This relation can be viewed as the convolution of the sequence $\{f_n\}$ with the finite impulse response $\{z_k^{-n}\}$, $0 \leqq n \leqq N-1$,

$$s_n(z_k) = \sum_{m=0}^{n} f_{n-m} z_k^{-m}. \tag{6.16b}$$

The $\mathscr{Z}$ transform of the sliding spectrum is

$$S_k(z) = F(z) \frac{z^N - z_k^{-N}}{z^{N-1}(z - z_k^{-1})}, \tag{6.17}$$

where $F(z) = \mathscr{Z}\{f_n\}$. Since $z_k^{-N} = 1$, we further have

$$S_k(z) = F(z) \frac{z}{z - z_k^{-1}} (1 - z^{-N}) = F(z)\, G_k(z)\, (1 - z^{-N}), \tag{6.18}$$

where $G_k(z) = z/(z - z_k^{-1})$. Comparing this result with (5.11) we see that $S_k(z)$ is the transform of the response of a digital filter constructed on the principle of frequency sampling with one recursive branch with transfer function $G_k(z)$, which has a complex pole at the point $z = z_k^{-1} = e^{-j(2\pi/N)k}$ on the unit circle. The general method of realization is indicated in Fig. 59.

The inverse transform of relation (6.18) yields an alternative representation of the sequence $\{s_n(z_k)\}$:

$$s_n(z_k) = h_{k,n} - h_{k,n-N}, \tag{6.19a}$$

where

$$h_{k,n} = \mathscr{Z}^{-1}\{H_k(z)\} = \mathscr{Z}^{-1}\{F(z)\, G_k(z)\}.$$

Putting $n = N$, we obtain

$$s_N(z_k) = h_{k,N} - h_{k,0}, \tag{6.19b}$$

where

$$h_{k,n} = \sum_{m=0}^{n} f_{n-m} z_k^{-m} \tag{6.20}$$

is formally identical with relation (6.16b). However, the difference consists in the fact that $\{z_k^{-n}\}$ is now the impulse response of a filter with infinite impulse response. To be able to apply relation (6.20) to the computation of the value of the complex spectrum we use n only from the interval $0 \leqq n \leqq N$ and we put $f_N = 0$. Thus, we have

$$s_N(z_k) = h_{k,N} = \sum_{m=0}^{N-1} f_n z_k^{-n} =: F_k. \tag{6.21}$$

Otherwise stated, we compute the response of a digital filter with transfer function $G_k(z)$ to the sequence $\{f_n\} = \{f_0, f_1, \ldots, f_{N-1}, 0\}$. The value of the response for $n = N$ is equal to the value of the discrete Fourier transform for $z_k = e^{j(2\pi/N)k}$.

Let us multiply the numerator as well as the denominator of $G_k(z)$ by the binomial $(z - z_k)$. We obtain a transfer function with real constants in the denominator:

$$G_k(z) = \frac{z(z - z_k)}{(z - z_k)(z - z_k^{-1})} = \frac{z^2 - z\left(\cos\frac{2\pi}{N}k + j\sin\frac{2\pi}{N}k\right)}{z^2 - 2\cos\frac{2\pi}{N}k \, . \, z + 1}. \tag{6.22}$$

Since we do not need the response of the filter $G_k(z)$ for all values of n, it is more convenient to work with the auxiliary transfer function

$$U_k(z) = \frac{z^2}{z^2 - 2\cos\frac{2\pi}{N}k \, . \, z + 1} \tag{6.23}$$

and to complete the computation of the response by the numerator till for $n = N$.

If we express the computation of the response of the filter $G_k(z)$ with the help of recurrent relations, we arrive at the familiar relations of the *Goertzel algorithm* for the evaluation of the DFT [29], [38]

$$u_{k,n} = f_n + 2\cos\frac{2\pi}{N}k \, . \, u_{k,n-1} - u_{k,n-2}, \quad n = 0, 1, \ldots, N, \tag{6.24a}$$

$$\operatorname{Re} F_k = \operatorname{Re} h_{k,N} = u_{k,N} - \cos\frac{2\pi}{N}k \, . \, u_{k,N-1}, \tag{6.24b}$$

$$\operatorname{Im} F_k = \operatorname{Im} h_{k,N} = \sin\frac{2\pi}{N}k \, . \, u_{k,n-1}, \tag{6.24c}$$

for the initial conditions $u_{k,-1} = u_{k,-2} = 0$ and for $f_N = 0$.

The forced condition $f_N = 0$ is induced by the presence of z^2 in the numerator of (6.22) and prevents the repeated application of the Goertzel algorithm for successive sections of an infinite sequence. Therefore, let us decompose the transfer function $G_k(z)$ in (6.22) into two terms as follows

$$G_k(z) = 1 + \frac{z\left(\cos\frac{2\pi}{N}k - \mathrm{j}\sin\frac{2\pi}{N}k\right) - 1}{z^2 - 2\cos\frac{2\pi}{N}k \,.\, z + 1} \tag{6.25}$$

and apply the same consideration as above. We obtain the modified Goertzel algorithm

$$u_{k,n} = f_n + 2\cos\frac{2\pi}{N}k \,.\, u_{k,n-1} - u_{k,n-2}\,, \qquad n = 0, 1, \ldots, (N-1)\,, \tag{6.25a}$$

$$\operatorname{Re} F_k = \operatorname{Re} h_{k,N} = \cos\frac{2\pi}{N}k \,.\, u_{k,N-1} - u_{k,N-2}\,, \tag{6.25b}$$

$$\operatorname{Im} F_k = \operatorname{Im} h_{k,N} = \sin\frac{2\pi}{N}k \,.\, u_{k,N-1}\,, \tag{6.25c}$$

with initial conditions $u_{k,-1} = u_{k,-2} = 0$.

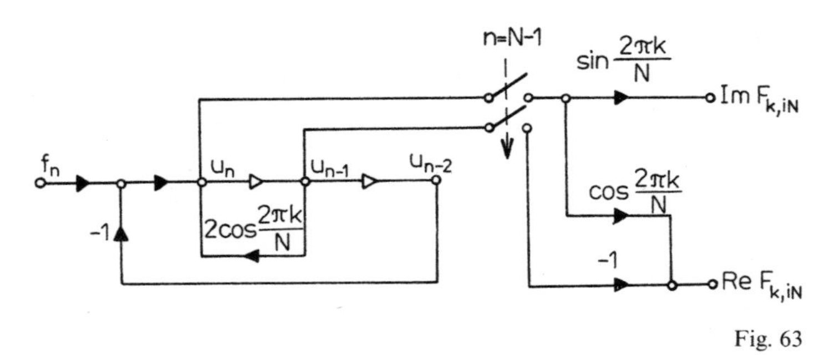

Fig. 63

The structure which implements this relation is shown in Fig. 63. This circuit enables to obtain sequentially the values $s_{iN}(z_k) = h_{k,iN} = F_{k,iN}$ for $i = 1, 2, \ldots$ without the forcible condition $f_{iN} = 0$. Consequently, it is possible to process a long sequence divided into sections of N samples and to obtain the decimated sequence of spectral values in jumps of size N. Before processing each section, for $n = iN$, $i = 1, 2, \ldots$, it is necessary always to satisfy the condition $u_{k,iN-1} = u_{k,iN-2} = 0$.

In [56] another modification of the Goertzel algorithm is presented which leads to a structure with smaller sensitivity to roundoff errors of the coefficients. The resulting modification is related to the Reintsch algorithm [38].

The discussed algorithms offer the possibility of determining the individual terms of the sequence of the discrete Fourier transform as functions of a decimated discrete time variable $n = iN$ by conversion to the problem of the filtering of the analyzed sequence by a simple filter of second order. In problems in which the entire complex spectrum, i.e. the complete sequence $\{F_0, F_1, \ldots, F_{N-1}\}$, is not desired, this procedure is simpler and leads to a smaller number of arithmetic operations than the algorithm of the fast Fourier transform which mostly requires that N be an integer power of the number 2.

For the computation of a point of the complex spectrum F_k two constants, $\cos(2\pi/N)\,k$ and $\sin(2\pi/N)\,k$, are necessary. This can be avoided if we make use of the fundamental property of digital filters which consists in their tuning by sampling frequency variation. This can be achieved practically by processing the sequence sampled with frequency $f_s' = kf_s$, $k > 1$, by a filter with coefficients for $k = 1$. For $n = iN$, $i = 1, 2, \ldots$, we then obtain at the output of the digital filter the values $s_{iN}(z_1) = h_{1,iN} = \operatorname{Re} F_k + \mathrm{j} \operatorname{Im} F_k$.

Finally, it is still possible to draw attention to the relation of the Goertzel algorithm and of its modifications and the Collatz scheme (or also the Horner scheme) applied in Subsection 5.1.3 to the computation of the frequency characteristics of digital filters. Comparing the computational operations we find out that the algorithms are identical although their derivations – and especially their physical interpretations – are different.

6.1.3. *Computation of the Frequency Characteristic of a System from Its Response to a Given Input Signal*

From the preceding subsections we are acquainted with several possibilities of the computation of the complex spectrum to a given discrete signal. Now, let us try to indicate briefly the procedure of the computation of the complex frequency characteristic of a system if its input and output signals are known. This problem is encountered in those applications where it is not possible to measure the frequency characteristic of the system directly, e.g. by a harmonic signal with variable frequency, but we have to compute it from its response to the operational signal.

Let us consider the system of Fig. 64 for which

$$H(z) = F(z)\,G(z)\,,$$

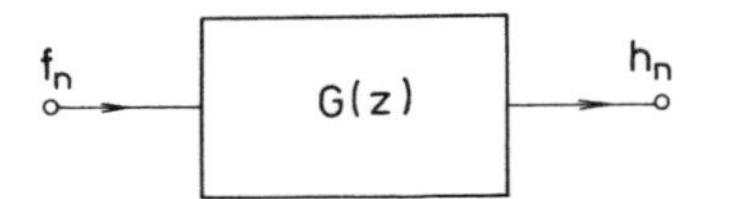

Fig. 64

where $F(z)$ and $H(z)$ are the transforms of the input and output signals, respectively. Obviously, the transfer function of the system is

$$G(z) = \frac{H(z)}{F(z)}. \tag{6.26}$$

The complex frequency response is obtained for $z = \mathrm{e}^{\mathrm{j}\omega T}$. This leads to the computation of the spectra of the output and input signals. If the input signal is represented by an analytic relation, the corresponding spectrum can be obtained directly from its $\mathscr{Z}$ transform. If, e.g., the sequence $\{h_n\}$ is constant starting from $n = N - 1$, i.e. if $h_n = h_{N-1}$ for $n \geqq N - 1$, its transform can be decomposed into two parts:

$$\begin{aligned}\mathscr{Z}\{h_n\} &= \sum_{n=0}^{N-1} h_n z^{-n} + h_{N-1} \sum_{n=N}^{+\infty} z^{-n} = \\ &= \sum_{n=0}^{N-1} h_n z^{-n} + h_{N-1} z^{-N} \sum_{n=0}^{+\infty} z^{-n} = \\ &= \sum_{n=0}^{N-1} h_n z^{-n} + h_{N-1} \frac{z^{-(N-1)}}{z-1}.\end{aligned} \tag{6.27}$$

The spectrum of the first part is determined, e.g., by the Collatz scheme or by FFT, the spectrum of the second part is obtained by evaluation for $z = \mathrm{e}^{\mathrm{j}\omega T}$.

In case that the input and output signals are infinite sequences, it is necessary to use their finite sections obtained and modified by a suitable window.

A further possibility of the determination of the complex frequency response of a system consists in the identification of the transfer function of the system from its response to a given input signal, this by the procedure presented in Subsections 4.1.5 or 6.2.3. The complex frequency response is then obtained from the transfer function by the numerical procedure of Subsection 5.1.3.

6.2. Restauration of the Input Signal of a System from Its Response

Let us return to Fig. 64. The response of the system to the input signal is given by the discrete convolution

$$h_n = f_n * g_n, \tag{6.28}$$

where $\{g_n\}$ is the impulse response of the system. Now, let us formulate the following problem: A system is given either by its transfer function $G(z)$ or by its impulse response $\{g_n\}$, we know the response $\{h_n\}$ and it is necessary to reconstruct the input signal $\{f_n\}$. Thus, we have to solve the inverse problem to the convolution. This problem is usually called the *deconvolution* of the response $\{h_n\}$ by the impulse response $\{g_n\}$. If we express the relationship (6.28) with the aid of the $\mathscr{Z}$ transform, it is possible to represent the deconvolution by the following relation which indicates already the process of the solution of the problem:

$$F(z) = H(z)\frac{1}{G(z)} = H(z)\, G_i(z), \tag{6.29}$$

where $G_i(z)$ is the *inverse transfer function.* The input signal $\{f_n\}$ is thus again given by the convolution

$$f_n = h_n * g_{i,n} \tag{6.30}$$

where, however, $\{g_{i,n}\} = \mathscr{Z}^{-1}\{G_i(z)\}$ is the impulse response of the inverse transfer function $G_i(z)$.

If $G(z) = P(z)/Q(z)$, where $P(z)$ and $Q(z)$ are polynomials in the variable z, then $G_i(z) = Q(z)/P(z)$. Consequently, the condition of stability for $G_i(z)$

is identical with the condition that $G(z)$ be the transfer function of a minimum phase system, i.e. that it have zeros inside the unit circle. Further, from the condition that the inverse system be causal it follows that the order of the numerator of the function $G(z)$ must be equal to the order of the denominator.

It is obvious that the mentioned conditions will be satisfied only in special cases, and even then the problem need not be easily solvable. If, for instance, the response $\{h_n\}$ contains disturbing components arising in the process of its registration, be they deterministic (e.g. the mean value and linear trend) or random (noise), we must either preprocess it by a suitable filter or correct the inverse transfer function $G_i(z)$ in the corresponding manner. In the opposite case, the disturbing components could be emphasized by deconvolution.

More frequently, we encounter problems which cannot be solved exactly, for which the procedure depends on the properties of the system, and which have to be solved approximately – either in the time domain or in the frequency domain. The homomorphic deconvolution of signals is a special case which enables, under certain assumptions, to decompose the response into the input signal and the impulse response of the system (see Subsection 6.2.4).

Before discussing some of the deconvolution methods we point out the connection of deconvolution and identification. The means of deconvolution are the inverse transfer function $G_i(z)$, or the impulse response $g_{i,n} = \mathscr{Z}^{-1}\{G_i(z)\}$. From relation (6.29) it is obvious that $G_i(z) = 1/G(z)$, whence we have

$$G(z)\, G_i(z) = 1 = \mathscr{Z}\{\delta_n\}\,, \tag{6.31}$$

where $\{\delta_n\}$ is the discrete unit impulse. It is possible to interpret this relation in accordance with Fig. 65 and to formulate the construction of the inverse

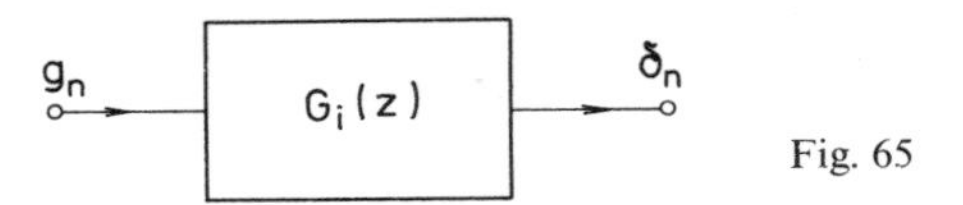

Fig. 65

transfer function as the problem of the identification of $G_i(z)$ from its response $\{\delta_n\}$ to the input signal $\{g_n\}$. This problem was treated in general in Subsection 4.1.5, and we shall still resume the discussion in Subsection 6.2.3.

6.2.1. *Deconvolution in the Time Domain*

In this subsection we shall complete the introductory considerations by several special methods of solving deconvolution problems. First, we discuss the case in which the transfer function $G(z)$ is of the nonminimum phase type, i.e. the inverse transfer function $G_i(z)$ is unstable in the sense of the one-sided $\mathscr{Z}$ transform. The transfer function can then be interpreted in accordance with Section 2.4 and Subsection 3.2.1 and the deconvolution may be performed two-sided.

Thus, let us have the inverse transfer function $G_i(z)$ which is regular in the region $|z| > 1/R > 1$, i.e. which represents an unstable system in the one-sided $\mathscr{Z}$ transform. If the transfer function is understood in the sense of the two-sided transform, it will represent a non-causal but stable system. For the inverse transfer function we introduce the notation $G_i(z) = G_{\mathrm{II}}(z)$ and decompose it into the product $G_{\mathrm{II}}(z) = G_+(z)\, G_-(z)$. Here, $G_+(z)$ is the causal part of transfer function $G_{\mathrm{II}}(z)$ and it is regular in the region $|z| \geqq 1$. On the other hand, $G_-(z)$ is the anticipative part of $G_{\mathrm{II}}(z)$ and it is regular for $|z| \leqq 1$. According to Subsection 2.4.2, the deconvolution will thus be performed twice. Expressed with the aid of the $\mathscr{Z}$ transform, we have

$$F_+(z) = H(z)\, G_+(z)\,, \tag{6.32a}$$

$$F(z) \;= F_{\mathrm{II}}(z) = F_+(z)\, G_-(z)\,. \tag{6.32b}$$

The first relation represents common causal filtering of the sequence $\{h_n\}$ by a filter with causal transfer function $G_+(z)$, the second represents anticipative filtering of the signal $\mathscr{Z}^{-1}\{F_+(z)\}$ from the end to the beginning. As the result we have the two-sided sequence $\{f_n\}$. It is obvious that only finite sequences can be processed in this manner. This approach is usually applied to the processing of records; the part of the signal which we are interested in must be located in the central part of the record. The reason is that the end of the resulting sequence $\{f_n\}$ is distorted by the causal transient of the filter with transfer function $G_+(z)$ while the beginning (leading values) of the sequence $\{f_n\}$ is distorted by the "anticipative transient" of the filter with transfer function $G_-(z)$ [52].

It is worth mentioning that this method of two-sided signal processing gives the possibility of *filtering with zero phase shift*. Namely, if the zeros and poles of the function $G_+(z)$ are located symmetrically, with respect to the unit circle, to the zeros and poles of the function $G_-(z)$, the resulting frequency characteristic of transfer function $G_{II}(z) = G_+(z)\, G_-(z)$ is real. This method of filtering is used with advantage in geophysics or in medicine, for instance, since it enables the application of filters with infinite impulse response, thus with a relatively small number of arithmetic operations for one response value, to obtain a phase undistorted response.

The same method of non-causal two-sided filtering can be used in case that the transfer function $G(z) = 1/G_i(z)$ is given by the $\mathscr{Z}$ transform of the finite impulse response, thus in the form [42]

$$G(z) = \sum_{n=0}^{N-1} g_n z^{-n} = \frac{1}{z^{N-1}} \sum_{n=0}^{N-1} g_n z^{N-1-n} . \tag{6.33}$$

If we solve the equation $G(z) = 0$ and classify the roots into roots lying inside and outside the unit circle, we easily decompose the inverse transfer function $G_i(z) = G_{II}(z) = 1/G(z)$ into the product of $G_+(z)$ and $G_-(z)$. Further, we proceed as in the previous case. This procedure is more difficult because it requires the factorization of a polynomial of high order ($N = 50 \div 300$, as a rule). However, it is solvable with the aid of a suitable algorithm. This approach will still be discussed in Subsection 6.2.4.

6.2.2. *Deconvolution in the Frequency Domain*

For the sake of completeness we mention briefly the problem of the restoration of the input signal of a system with the aid of its complex frequency characteristic and the complex spectrum of its response $\{h_n\}$.

In relation (6.29) let us put $z = e^{j\omega T}$. For the Fourier transform (spectrum) of the input signal $\{f_n\}$ we have

$$F(e^{j\omega T}) = H(e^{j\omega T}) \frac{1}{G(e^{j\omega T})} = H(e^{j\omega T})\, G_i(e^{j\omega T}) . \tag{6.34}$$

If we represent $G(e^{j\omega T})$ in polar form

$$G(e^{j\omega T}) = A(\omega)\, e^{j\varphi(\omega)} ,$$

we obtain for the frequency characteristic of the inverse transfer function the relation

$$G_i(e^{j\omega T}) = \frac{1}{A(\omega)} e^{-j\varphi(\omega)} . \tag{6.35}$$

The sequence of values of the input signal is formally obtained, in line with Subsection 3.2.2, by the inverse Fourier transform

$$f_n = \frac{1}{2\pi T} \int_0^{2\pi/T} F(e^{j\omega T})\, e^{j\omega T}\, d\omega =$$

$$= \frac{1}{2\pi T} \int_0^{2\pi/T} H(e^{j\omega T})\, G_i(e^{j\omega T})\, e^{j\omega T}\, d\omega\,. \tag{6.36}$$

For practical reasons we are forced to work with finite sequences (possibly with the exception of the function $G(z)$ which can be given in compact form by the transfer function). For this reason we apply, with advantage, the discrete Fourier transform to the computation of the Fourier transforms. Then it is possible to summarize the deconvolution procedure in the frequency domain into the following relations:

$$\tilde{f}_n = \frac{1}{N} \sum_{k=0}^{N-1} H_k G_{i,k}\, e^{j(2\pi/T)kn}\,, \tag{6.36a}$$

$$H_k = H(z)\Big|_{z = e^{j(2\pi/N)k}}\,, \tag{6.36b}$$

$$G_{i,k} = G_i(z)\Big|_{z = e^{j(2\pi/N)k}} = \frac{1}{G_k} = G^{-1}(z)\Big|_{z = e^{j(2\pi/N)k}} =$$

$$= \frac{1}{|G_k|}\, e^{-j \arg G_k}\,. \tag{6.37}$$

Now, it is obvious that $\{f_n\}$ will be a periodic sequence with period N in consequence of the application of DFT.

6.2.3. *Application of Linear Prediction to the Construction of the Inverse Transfer Function*

Now, let us devote our attention to a special case of the construction of the inverse transfer function. The approach is useful in a number of applications, e.g. in speech analysis, geophysics, and also in spectral analysis. Taking into account the continuity of the exposition we derive the method in the time domain. Similar results could also be obtained under the assumption that we are looking for the inverse transfer function to a finite realization of a random signal. Methods which are based on statistical considerations are called in the literature maximum likelihood methods and maximum entropy methods [9], [25].

We return to relation (6.31) which corresponds to Fig. 65,

$$G(z)\,G_i(z) = \mathscr{Z}\{\delta_n\}.$$

Here, $G(z) = \mathscr{Z}\{g_n\}$ is the transform of the impulse response of the system. We introduce the condition that the inverse transfer function should be a polynomial in the variable z^{-1}:

$$G_i(z) = 1 - \sum_{i=1}^{r} a_i z^{-i} = 1 - A(z). \tag{6.38}$$

This form of the inverse transfer function is convenient for several applications since it corresponds to the physical models of some systems, e.g. the model of speech analysis. Besides, this form leads to a reasonable simplification of the computational algorithm. If we express the transform $G(z)$ from (6.31) with the aid of the inverse transfer function, we obtain the relation

$$G(z) = \frac{\mathscr{Z}\{\delta_n\}}{G_i(z)} = \frac{\mathscr{Z}\{\delta_n\}}{1 - A(z)} \tag{6.39}$$

which corresponds, e.g., to the model of speech generation, and which represents the function $G(z)$ in compact form. The inverse transform of relation (6.31) yields the response of the inverse filter to the input signal $\{g_n\}$ represented by the impulse response $\{a_n\} = \mathscr{Z}^{-1}\{A(z)\}$

$$\delta_n = g_n - \sum_{i=1}^{r} a_i g_{n-i}, \tag{6.40}$$

which can be represented by the system shown in Fig. 66. The response

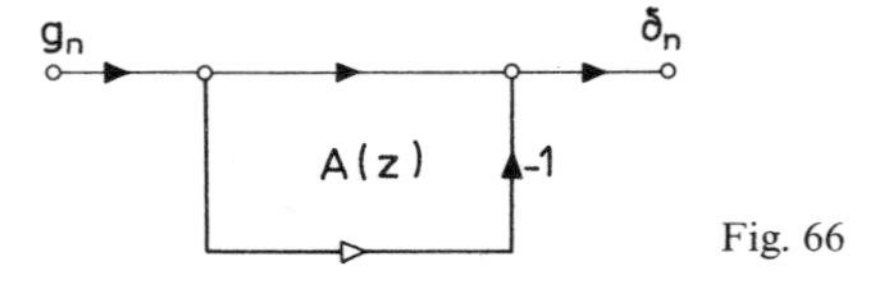

Fig. 66

of the filter with transfer function $A(z)$ to the sequence $\{g_n\}$ is denoted by

$$\tilde{g}_n = \sum_{i=1}^{r} a_i g_{n-i}.$$

Let us consider the response to be a sequence predicted from the last r terms of the sequence $\{g_n\}$. For this reason the filter is also called a *prediction filter*. The *prediction error* is given by the difference

$$e_n = g_n - \tilde{g}_n = g_n - \sum_{i=1}^{r} a_i g_{n-i}\,. \tag{6.41}$$

Comparison with (6.40) shows that in the case when $\{g_n\}$ is the response of a filter of type (6.39) we have $e_n = \delta_n$ for the prediction error. The coefficients of the prediction filter $\{a_i\}$ will be determined from the condition that the energy of the error $\{e_n\}$, defined by the relation

$$E = \sum_{n=0}^{+\infty} e_n^2 = \sum_{n=0}^{+\infty} \left(g_n - \sum_{i=1}^{r} a_i g_{n-i} \right)^2, \tag{6.42}$$

should be minimal. The coefficients a_k which minimize the sum of squared errors must satisfy the equations $\partial E / \partial a_k = 0$, for $k = 1, 2, \ldots, r$. Upon differentiation this yields the following system of linear equations for the coefficients a_k:

$$- \sum_{n=0}^{+\infty} g_n g_{n-k} + \sum_{i=1}^{r} a_i \sum_{n=0}^{+\infty} g_{n-i} g_{n-k} = 0\,. \tag{6.43}$$

Exchanging the subscripts and assuming that the sequence $\{g_n\}$ is one-sided, i.e. if $g_n = 0$ for $n < 0$, we obtain the normal system of equations

$$\sum_{i=1}^{r} a_i \sum_{n=0}^{+\infty} g_n g_{n+i-k} = \sum_{n=0}^{+\infty} g_n g_{n-k}\,. \tag{6.44}$$

The infinite sums represent *aperiodic autocorrelation sequences* (see Subsection 2.3.1) for which we have

$$\sum_{n=0}^{+\infty} g_n g_{n-m} = \psi_m = \psi_{-m} = \psi_{|m|}\,. \tag{6.45a}$$

In the common case we deal with a finite sequence $\{g_n\}$ for which $g_n = 0$ for $n > N - 1$. Then the autocorrelation sequence is

$$\psi_m = \sum_{n=0}^{+\infty} g_n g_{n-m} = \sum_{n=0}^{N-1-m} g_n g_{n+m} = \psi_{|m|}\,. \tag{6.45b}$$

The normal system of r equations in r unknowns can then be written in the clearer form

$$\sum_{i=1}^{r} a_i \psi_{i-k} = \psi_k, \qquad k = 1, 2, \ldots, r \tag{6.46}$$

or, in matrix notation,

$$\boldsymbol{\phi a} = \boldsymbol{\psi}, \tag{6.47}$$

where

$$\boldsymbol{\phi} = \begin{bmatrix} \psi_0 & \psi_1 & \psi_2 & \cdots & \psi_{r-1} \\ \psi_1 & \psi_0 & \psi_1 & \cdots & \psi_{r-2} \\ \psi_2 & \psi_1 & \psi_0 & \cdots & \psi_{r-3} \\ \vdots & \vdots & \vdots & & \vdots \\ \psi_{r-1} & \psi_{r-2} & \psi_{r-3} & \cdots & \psi_0 \end{bmatrix},$$

$${}^T\boldsymbol{a} = [a_1, a_2, \ldots, a_r],$$

$${}^T\boldsymbol{\psi} = [\psi_1, \psi_2, \ldots, \psi_r].$$

Since the matrix $\boldsymbol{\phi}$ is formed by the terms of the autocorrelation sequence, the method of the determination of the coefficients a_i is usually called the *autocorrelation method*. The matrix $\boldsymbol{\phi}$ is symmetrical, positive definite, and has identical elements along the main diagonal. Matrices of this type are called *Toeplitz matrices*. For the solution of the system of equations (6.47) there exist fast and effective algorithms, e.g. the Levinson algorithm [25], [33].

Substitution of (6.45) into relation (6.42) yields the energy of the prediction error represented with the aid of the autocorrelation sequence and the coefficients of the prediction filter

$$E = \sum_{n=0}^{+\infty} e_n^2 = \sum_{n=0}^{+\infty} g_n^2 - \sum_{i=1}^{r} a_i \sum_{n=0}^{+\infty} g_n g_{n-i} = \psi_0 - \sum_{i=1}^{r} a_i \psi_i. \tag{6.48}$$

For the sake of completeness we still have to mention a further possible method of the computation of the infinite sums from relation (6.45). Namely, if we accumulate a constant number of products when computing ψ_m, the matrix of the system will be symmetrical but not Toeplitz. For reasons of tradition this approach is usually labelled in the literature the covariance method for the construction of the inverse filter. The details may be found in [9], [25], [33]. The method of system identification presented in Subsection 4.1.5 can therefore be called the covariance method.

6.2.4. *Homomorphic Filtering of Signals*

Homomorphic filtering is a *nonlinear method of signal processing* which is based on the generalized principle of superposition and makes use of linear systems methods. It is applied to processing of signals which are combined by multiplication or convolution.

The first work in this field was published by Bogert, Healy and Tukey [4] in 1963. They analyzed a signal consisting of the sum of two signals, a primary and a secondary signal, where the secondary signal originated by delay of the primary signal and its modification by a multiplicative constant. Such a signal it is possible to encounter, e.g., in acoustic measurements, in speech and seismic signal processing, in EEG analysis, etc. The logarithm of the power spectrum of such a signal is rippled, whereas the length of the period of the periodic component is the reciprocal value of the delay between the primary and the secondary signal. The "frequency" of the periodic component can be determined by spectral analysis of the logarithm of the power spectrum. The delay shows itself by a "resonance peak" in this analysis. Since the "frequency" of the periodic component has the time dimension, the mentioned authors have introduced the notion of cepstrum for the spectrum of the logarithm of the power spectrum, as a paraphrasis of the word spectrum.

In the years which followed, Oppenheim [13], [29], [31] completed the method and elaborated it in detail. He has introduced the notion of the complex cepstrum of a sequence. Contrary to the cepstrum introduced by Bogert *et al.*, the complex cepstrum includes also information concerning the phase of the signal. Thus, it makes it also possible to reconstruct the signal and perform deconvolution.

The separation of signals by way of cepstral analysis – homomorphic filtering – may be formulated with advantage and in compact form with the aid of the $\mathscr{Z}$ transform. The relationship between the $\mathscr{Z}$ transform and the discrete Fourier transform with the FFT algorithm, yields also effective computational procedures.

The *complex cepstrum* of the real sequence $\{x_n\}$, $n = 1, 2, \ldots$, is formally defined using the complex logarithmic function

$$\hat{x}_n = \mathscr{Z}^{-1}\{\hat{X}(z)\} = \mathscr{Z}^{-1}\{\ln X(z)\}, \tag{6.49}$$

where $X(z) = \mathscr{Z}\{x_n\}$. The function $\hat{X}(z) = \ln X(z)$ is called the *transform of the complex cepstrum.* Since $\hat{X}(z)$ may possess singular points in the entire complex plane, we have to understand the inverse $\mathscr{Z}$ transform in the sense of the two-sided $\mathscr{Z}$ transform. According to Theorem 22, the existence of $\{\hat{x}_n\}$ is then given by the condition that the function

$$\hat{X}(z) = \hat{X}_{\mathrm{II}}(z) = \ln X(z) = \ln \mathscr{Z}\{x_n\} \tag{6.50}$$

is regular in the annulus $1/R_+ < |z| < R_-$. Then we have $\hat{X}_{\mathrm{II}}(z) = \mathscr{Z}_{\mathrm{II}}\{\hat{x}_n\}$, where

$$\hat{x}_n = \frac{1}{2\pi \mathrm{j}} \oint_C \hat{X}_{\mathrm{II}}(z)\, z^{n-1}\, \mathrm{d}z \qquad \text{for} \qquad n = \ldots, -2, -1, 0, 1, 2, \ldots. \tag{6.51}$$

We integrate along the circle C for which $z = \varrho \mathrm{e}^{\mathrm{j}\varphi}$, $1/R_+ < \varrho < R_-$ and $0 \leqq \varphi \leqq 2\pi$.

Relation (6.51) may be called the *cepstral transform of the sequence* $\{x_n\}$ and we write symbolically

$$\hat{x}_n = \mathscr{K}\{x_n\} = \mathscr{Z}^{-1}\{\ln \mathscr{Z}\{x_n\}\}.$$

The *inverse cepstral transform* follows from (6.50). Symbolically, it is given by

$$x_n = \mathscr{K}^{-1}\{\hat{x}_n\} = \mathscr{Z}^{-1}\{\mathrm{e}^{\mathscr{Z}\{x_n\}}\}. \tag{6.52}$$

From relation (6.51) it follows that the complex cepstrum $\{\hat{x}_n\}$ is generally a two-sided sequence which it is possible to interpret as the impulse response of a non-causal system with transfer function $\hat{X}_{\mathrm{II}}(z)$. For practical reasons it is necessary that this non-causal system be stable. This leads to the requirement that $\hat{X}_{\mathrm{II}}(z)$ be a regular function in an annular region which includes the unit circle. Relation (6.50) implies that this requirement will be satisfied only if not only the poles but also the zeros of the transform $X(z)$ will not lie on the unit circle.

The transform of the complex cepstrum $\hat{X}_{\mathrm{II}}(z)$ may be decomposed into two one-sided transforms

$$\hat{X}_{\mathrm{II}}(z) = \hat{X}_+(z) + \hat{X}_-(z),$$

where

$$\hat{X}_+(z) = \sum_{n=0}^{+\infty} \hat{x}_n z^{-n} \qquad \text{and} \qquad \hat{X}_-(z) = \sum_{n=-1}^{+\infty} \hat{x}_n z^{-n}.$$

The transform $\hat{X}_+(z)$ is a regular function in the region $|z| \geqq 1 \geqq 1/R_+$, the transform $\hat{X}_-(z)$ is a regular function in the region $|z| < R_-$. In the sense of Subsection 3.2.1, $\{\hat{x}_n\}$ for $n \geqq 0$ is then called the *causal part of the complex cepstrum* while $\{\hat{x}_n\}$ for $n < 0$ is called the *anticipative part of the complex cepstrum.*

For completeness, we still mention the relationship between the complex cepstrum $\{\hat{x}_n\}$ and the cepstrum defined in the sense of Bogert *et al.* This cepstrum is also sometimes called the *power cepstrum* or the *real cepstrum* and it is denoted by $\{c_n\}$. With the aid of the $\mathscr{Z}$ transform it is possible to introduce the cepstrum of the sequence $\{x_n\}$ by the relation

$$c_n = \frac{1}{2} \mathscr{Z}^{-1}\{\ln |X(z)|^2\} = \frac{1}{2} \frac{1}{2\pi \mathrm{j}} \oint_C \ln \left[X\left(\frac{1}{z}\right) X(z) \right] z^{n-1} \, \mathrm{d}z \,, \tag{6.53}$$

where we integrate along the unit circle. The function $X(1/z)\, X(z) = \Psi(z)$ yields for $z = \mathrm{e}^{\mathrm{j}\omega T}$ the power spectral density which is always a real non-negative function of the radian frequency ω. Consequently, it is possible to work with the real logarithm in the definition of the cepstrum. By the Wiener–Khintchine theorem (see Subsection 2.3.1) the cepstrum $\{c_n\}$ is expressed by the autocorrelation sequence $\{\psi_n\}$ as follows

$$c_n = \frac{1}{2} \mathscr{Z}^{-1}\{\ln \Psi(z)\} = \frac{1}{2} \mathscr{Z}^{-1}\{\ln \mathscr{Z}\{\psi_n\}\}\,, \tag{6.54}$$

where $\Psi(z) = X(z)\, X(1/z)$. Upon a simple modification of equation (6.53) we reach the following relation between the cepstrum and the complex cepstrum:

$$c_n = \frac{\hat{x}_n + \hat{x}_{-n}}{2}. \tag{6.55}$$

Since $c_n = c_{-n}$, it suffices to limit our considerations to the branch of the cepstrum for $n \geqq 0$.

In what follows, we demonstrate on a simple example the application of cepstral analysis to the deconvolution of signals.

Let us have a discrete signal $\{x_n\}$ for which

$$x_n = f_n + af_{n-k}. \tag{6.56}$$

The sequence $\{f_n\}$, $f_n = 0$ for $n < 0$, is the primary signal, the sequence $\{af_{n-k}\}$ is the secondary signal given by the delay of signal $\{f_n\}$ by k sampling intervals, with the multiplicative constant $0 < |a| < 1$. For the transform of $\{x_n\}$ we have

$$X(z) = F(z)(1 + az^{-k}) = F(z)\,G(z), \tag{6.57}$$

where $F(z) = \mathscr{Z}\{f_n\}$ and $G(z) = 1 + az^{-k}$. The signal $\{x_n\}$ can now be interpreted as the response of a system with transfer function $G(z)$ to the input signal $\{f_n\}$.

By (6.50), the transform of the complex cepstrum of the sequence $\{x_n\}$ is

$$\begin{aligned} \hat{X}(z) &= \mathscr{Z}\{\hat{x}_n\} = \ln X(z) = \ln F(z) + \ln G(z) = \\ &= \hat{F}(z) + \hat{G}(z) = \mathscr{Z}\{\hat{f}_n\} + \mathscr{Z}\{\hat{g}_n\}, \end{aligned} \tag{6.58a}$$

whence the inverse transform yields

$$\hat{x}_n = \hat{f}_n + \hat{g}_n. \tag{6.58b}$$

We see that the *complex cepstrum of the signal* $\{x_n\}$ *formed by the convolution of the primary signal* $\{f_n\}$ *and the impulse response* $\{g_n\}$ *is given by the sum of the corresponding complex cepstra.*

The computation of the complex cepstrum $\{\hat{f}_n\}$ will be discussed at the end of this subsection; now we concentrate our attention upon the investigation of the properties of the complex cepstrum of the impulse response.

With the aid of the list of transforms we obtain

$$\begin{aligned} \hat{G}(z) &= \ln(1 + az^{-k}) = -\ln\frac{z^k}{z^k + a} = \\ &= -\sum_{n=1}^{+\infty}\frac{(-a)^n}{n}z^{-kn} = -\sum_{m=0}^{+\infty}\hat{g}_m z^{-m}. \end{aligned} \tag{6.59a}$$

It is obvious that $\{\hat{g}_m\}$ is an infinite one-sided sequence. If we interpret the infinite sum in the sense of Theorem 7a on the change of the subscript of a sequence, we have

$$\hat{g}_m = -\frac{(-a)^n}{n} \quad \text{for} \quad m = kn\,, \quad n = 1, 2, 3, \dots,$$

$$\hat{g}_m = 0 \quad \text{for} \quad m \neq kn\,, \quad m = 0, 1, 2, \dots. \tag{6.59b}$$

Under the assumption that the complex cepstra do not overlap, or – otherwise stated – that $\hat{f}_n = 0$ for $n \geqq k$, we easily determine the complex cepstrum of $\{\hat{f}_n\}$ by weighting the sequence $\{\hat{x}_n\}$ by the cepstral window $\{w_n\}$. For this purpose we choose

$$w_n = +1 \quad \text{for} \quad n < k\,,$$

$$w_n = 0 \quad \text{for} \quad n \geqq k\,.$$

Thus we obtain $\hat{f}_n = w_n \hat{x}_n$. Hence, the inverse cepstral transform according to (6.52) yields the primary signal

$$f_n = \mathscr{K}^{-1}\{\hat{f}_n\} = \mathscr{Z}^{-1}\{e^{\mathscr{Z}\{\hat{f}_n\}}\} = \mathscr{Z}^{-1}\{e^{\hat{F}(z)}\}\,. \tag{6.60}$$

With the aid of the thus obtained complex cepstrum of the primary signal we easily determine, from relation (6.58b), the complex cepstrum $\{\hat{g}_n\}$. From $\{\hat{g}_n\}$, the impulse response $\{g_n\}$ is again determined by the inverse cepstral transform

$$g_n = \mathscr{K}^{-1}\{\hat{g}_n\} = \mathscr{K}^{-1}\{\hat{x}_n - \hat{f}_n\} = \mathscr{Z}^{-1}\{e^{\hat{X}(z) - \hat{F}(z)}\}\,. \tag{6.61}$$

Till now we have discussed the introduction of the complex cepstrum with the aid of the $\mathscr{Z}$ transform, the investigation of its basic properties, and using a simple example we have shown the procedure of homomorphic filtering for the deconvolution of signals. Now we shall devote our attention to methods of computing the complex cepstrum for the common case of finite sequences. In contradistinction to the method of cepstral analysis which is currently found in the literature based on the repeated computation of complex spectra with the aid of the fast Fourier transform, we first present two approaches based on the $\mathscr{Z}$ transform. These approaches will allow a different view of the relationship of the signal and the complex cepstrum. Further, they will enable the evaluation of different modifications of the signal from the point of view of their influence on the complex cepstrum. Finally, we will briefly outline the method of cepstral analysis based on the complex spectrum and on the application of DFT.

(a) *Computation of the complex cepstrum in closed form.*

Let us assume that a finite sequence $\{x_n\}$ of N terms is given which originated, e.g., from an infinite signal upon application of a suitable window. For the transform of the signal $\{x_n\}$ let us have

$$X(z) = \mathscr{Z}\{x_n\} = \sum_{n=0}^{N-1} x_n z^{-n} = x_0 \prod_{i=1}^{N-1} (1 - z_i z^{-1}), \tag{6.62a}$$

where z_i are zeros of the equation $X(z) = 0$. If the zeros z_i are divided into zeros lying inside and outside the unit circle (assume that no zero lies directly on the circle), then the transform $X(z)$ can be written in a form suitable for cepstral transformation according to relation (6.51)

$$X(z) = Az^{-S} \prod_{r=1}^{R} (1 - a_r z^{-1}) \prod_{s=1}^{S} (1 - b_s z), \tag{6.62b}$$

where R is the number of zeros for which $|a_r| < 1$, S is the number of zeros for which $b_s^{-1} > 1$, $R + S = N - 1$, and $A = x_0 \prod_{s=1}^{S} (-b_s)$. For real sequences the constant A is real. The term z^{-S} corresponds to the delay of the sequence $\{x_n\}$ by S sampling intervals. By a suitable modification of the sequence it is possible to achieve, without loss of generality, that $A > 0$, and that the term z^{-S} is compensated by a left shift of the sequence. In the considerations below, we will therefore start from the transform $X(z)$ in the form

$$X(z) = |A| \prod_{r=1}^{R} (1 - a_r z^{-1}) \prod_{s=1}^{S} (1 - b_s z). \tag{6.62c}$$

Taking the logarithm of (6.62b) we obtain the transform of the complex cepstrum

$$\hat{X}(z) = \ln |A| + \sum_{r=1}^{R} \ln (1 - a_r z^{-1}) + \sum_{s=1}^{S} \ln (1 - b_s z). \tag{6.63}$$

With the aid of the list of transforms and of a modification similar as in relation (6.59), we obtain the complex cepstrum of the finite sequence as follows:

$$\hat{x}_n = \ln|A| = \ln\left|x_0 \prod_{s=1}^{S}(-b_s)\right| \quad \text{for} \quad n = 0,$$

$$\hat{x}_n = -\prod_{r=1}^{R} \frac{a_r^n}{n} \quad \text{for} \quad n > 0 \quad \text{– the causal part,} \tag{6.64}$$

$$\hat{x}_n = \prod_{s=1}^{S} \frac{b_s^n}{n} \quad \text{for} \quad n < 0 \quad \text{– the anticipative part.}$$

Now, we summarize some basic facts which follow from the just derived relations:

(1) If the transform $X(z)$ has zeros only inside the unit circle, i.e. if it is of the *minimum-phase type*, the complex cepstrum is constituted only by the branch for $n \geqq 0$, i.e. it is *causal.*

(2) On the other hand, if $X(z)$ has zeros only outside the unit circle, i.e. if it is of the *maximum-phase type*, the corresponding complex cepstrum is also one-sided, but it is given by the branch for $n \leqq 0$, i.e. it is *anticipative.*

(3) If the transform $X(z)$ has zeros both inside and outside the unit circle, the corresponding complex cepstrum is non-causal, i.e. it has the causal as well as the anticipative part.

The presented computation method for the complex cepstrum assumes that the zeros of the transform $X(z)$ of the finite sequence are known. Since mostly only the sequence is known, the method asks for the solution of polynomial higher order equations. The solution of equations of order 200 till 300 may seem to be difficult, but the problem can be solved with the aid of suitable algorithms [37], [57]. The knowledge of the zeros permits, besides a simple algebraic computation procedure for the complex cepstra according to (6.64), to draw also conclusions concerning the properties of the signal and to exploit them eventually to modify the method. For instance, if any of the zeros lies on the unit circle, the fundamental condition of the cepstral transformation is not satisfied. In this case, it is possible to shift all the zeros in accordance with the theorem on the similarity of transforms. Namely, we have

$$\mathscr{Z}\{\alpha^n x_n\} = X(\alpha^{-1}z) = x_0 \prod_{i=1}^{N-1}(1 - \alpha z_i z^{-1}),$$

where α is a suitably chosen number. In the literature this approach is usually called the *exponential weighting* of a sequence and it is frequently used for the transformation of a non-causal complex cepstrum to a causal one.

(b) *Recurrent computation of the complex cepstrum.*

Now let us pay attention to the case when the sequence $\{x_n\}$ is known while the zeros of the corresponding transform are not known. By (6.50), the transform of the complex cepstrum is $\hat{X}(z) = \ln X(z)$. In the region of regularity of $X(z)$ we have

$$\frac{\mathrm{d}\hat{X}(z)}{\mathrm{d}z} = \frac{1}{X(z)} \frac{\mathrm{d}X(z)}{\mathrm{d}z}. \tag{6.65}$$

However, by Theorem 12 on the derivative of the transform (see Subsection 2.1.8) we also have

$$\frac{\mathrm{d}X(z)}{\mathrm{d}z} = -z^{-1} \mathscr{Z}\{nx_n\},$$

$$\frac{\mathrm{d}\hat{X}(z)}{\mathrm{d}z} = -z^{-1} \mathscr{Z}\{n\hat{x}_n\}. \tag{6.66}$$

If we substitute these relations into (6.65), we obtain upon performing the inverse transform (under the assumption of the one-sidedness of $\{x_n\}$ and $\{\hat{x}_n\}$, i.e. for $x_n = \hat{x}_n = 0$ for $n < 0$) the implicit relation

$$\sum_{k=0}^{n} k\hat{x}_n x_{n-k} = nx_n. \tag{6.67}$$

The requirement of one-sidedness corresponds to the condition that the transform $X(z)$ has zeros inside the unit circle only. Solving (6.67) for $\hat{x}_n$ we obtain the following recurrent relation for the *causal* complex cepstrum

$$\hat{x}_n = \frac{1}{x_0}\left(x_n - \frac{1}{n}\sum_{k=1}^{n-1} k\hat{x}_n x_{n-k}\right) \quad \text{for} \quad n > 0,$$

$$\hat{x}_n = 0 \quad \text{for} \quad n < 0. \tag{6.68}$$

As follows from Theorem 17 on the initial value of the object function we have $\hat{x}_0 = \ln |x_0|$.

Under the assumption that $x_n = \hat{x}_n = 0$ for $n > 0$ (which corresponds to the condition that the transform $X(z)$ has zeros outside the unit circle only), we analogously obtain the following relations for the *anticipative* complex cepstrum

$$\hat{x}_n = \frac{1}{x_0}\left(x_n - \frac{1}{n}\sum_{k=-1}^{n+1} k\hat{x}_k x_{n-k}\right) \quad \text{for} \quad n < 0,$$

$$\hat{x}_n = 0 \quad \text{for} \quad n > 0,$$

$$\hat{x}_0 = \ln|x_0|. \tag{6.69}$$

From relations (6.68) and (6.69) we can also obtain, under the appropriate conditions, the recurrent representation of the sequence $\{x_k\}$ with the aid of $\{\hat{x}_n\}$. This representation may be applied to the inverse cepstral transform.

(c) *Computation of the complex cepstrum with the aid of spectral analysis.*

The computation procedure of the complex cepstrum with the aid of spectral analysis follows from the relationship of the $\mathscr{Z}$ transform and the Fourier transform of a sequence. We arrive at this procedure with the aid of relations (6.50) and (6.51) in which we put $z = e^{j\omega T}$:

$$\hat{X}(e^{j\omega T}) = \ln X(e^{j\omega T}) = \ln \sum_{n=0}^{N-1} x_n e^{-j\omega T} =$$

$$= \ln|X(e^{j\omega T})| + j \arg X(e^{j\omega T}),$$

$$\hat{x}_n = \frac{1}{2\pi T}\int_0^{2\pi/T} \hat{X}(e^{j\omega T})\, e^{j\omega T}\, d\omega. \tag{6.70}$$

The computation of the complex cepstrum now asks (instead of the regularity of the function $\ln X(z)$ on the unit circle) for the continuity of the function $\hat{X}(e^{j\omega T})$ with respect to the variable ω. This condition must be satisfied by the real part as well as by the imaginary part of the function $\hat{X}(e^{j\omega T})$. The real part satisfies the condition under the assumption that $X(z)$ has no zero on the unit circle. However, the continuity of the imaginary part depends moreover on the method of computation of $\arg X(e^{j\omega T})$. The usual computation procedure based on the principal value of the argument yields a discontinuous curve since the principal value of the argument is defined only in the interval $-\pi < \arg X(e^{j\omega T}) \leqq \pi$. To satisfy the assumption of continuity we have to define the function $\ln X(e^{j\omega T})$ with the aid of the Riemann surface, i.e. we have to *unwrap* the function $\arg X(e^{j\omega T})$ applying a suitable algorithm [43]. A second possibility consists in the computation of the continuous function $\arg X(e^{j\omega T})$ by integrating its derivative, see Subsection 6.1.1. In both cases the function $\arg X(e^{j\omega T})$ has to be reduced to an odd function in the interval $[-\pi/T, \pi/T]$ by subtraction of the linear trend of the function $\arg X(e^{j\omega T})$. The trend corresponds to the number of zeros S of the transform $X(z)$ outside the unit circle (see the procedure ad (a)).

The computation of the direct and the inverse cepstral transforms is, as a rule, realized approximately with the aid of the discrete Fourier transform. For $\omega_k = 2\pi k/N$ we then have for the complex cepstrum the relations

$$\hat{X}_k = \ln X_k = \ln \sum_{n=0}^{N-1} x_n e^{j(2\pi/N)kn} = \ln |X_k| + j \arg X_k ,$$

$$\tilde{\hat{x}}_n \doteq \frac{1}{N} \sum_{k=0}^{N-1} \hat{X}_k e^{j(2\pi/N)kn} , \tag{6.71a}$$

while the inverse cepstral transform is given by the relation

$$\tilde{x}_n = \frac{1}{N} \sum_{k=0}^{N-1} e^{\hat{X}_k} e^{j(2\pi/N)kn} . \tag{6.71b}$$

In consequence of the application of DFT, the complex cepstrum $\{\tilde{\hat{x}}_n\}$ is a periodic sequence which is bound with the complex cepstrum $\{\hat{x}_n\}$ by the relation

$$\tilde{\hat{x}}_n = \sum_{i=-\infty}^{+\infty} \hat{x}_{n+iN} \tag{6.72}$$

which results in *aliasing in the cepstral domain.*

In all, three methods of the cepstral transform of sequences were presented. The last which makes use of DFT is relatively easy to perform thanks to the FFT algorithm, although its application asks for some cautiousness. The reader will find experience and a number of applications in [8], [25], [29], [31], [35].

Chapter 7

Appendices

7.1. Properties of the $\mathscr{Z}$ Transform

TABLE 7.1 Basic Relations – Correspondences and Limit Relations

Relation	Object function	Transform
Definition	$\{f_n\}$	$F(z) = \sum_{n=0}^{+\infty} f_n z^{-n}$
Linearity	$\left\{\sum_{i=0}^{l} c_i f_{i,n}\right\}$	$\sum_{i=0}^{l} c_i F_i(z)$
Left translation	$\{f_{n+k}\}$	$z^k \left[F(z) - \sum_{n=0}^{k-1} f_n z^{-n}\right]$
Right translation	$\{f_{n-k}\}$	$z^{-k} F(z)$
Change of subscript	$f_m = g_n, \quad m = ni$ $f_m = 0, \quad m \neq ni$	$F(z) = G(z^i); \qquad G(z) = F(z^{1/i})$
Similarity	$\{\lambda^n f_n\}$	$F\left(\frac{z}{\lambda}\right)$
Convolution	$\left\{\sum_{k=0}^{n} f_k g_{n-k}\right\}$	$F(z)\, G(z)$
Difference	$\Delta f_n = f_{n+1} - f_n$ $\Delta^k f_n = \Delta^{k-1} f_{n+1} - \Delta^k f_n$	$(z-1)\, F(z) - z f_0$ $(z-1)^k F(z) - z \sum_{i=0}^{k-1} (z-1)^{k-i} \Delta f_0$
Partial summation	$\left\{\sum_{k=0}^{n-1} f_k\right\}$	$\frac{1}{z-1} F(z)$
Differentiation of transform	$\{n f_n\}$	$-z \frac{\mathrm{d}}{\mathrm{d}z} F(z)$

TABLE 7.1 – *continued*

Relation	Object function	Transform
Integration of transform	$\left\{\frac{f_n}{n}\right\}$	$\int_z^{+\infty} F(\zeta)\,\zeta^{-1}\,\mathrm{d}\zeta$
Differentiation with respect to a parameter	$\left\{\frac{\mathrm{d}}{\mathrm{d}\xi} f_n(\xi)\right\}$	$\frac{\mathrm{d}}{\mathrm{d}\xi} F(z, \xi)$
Integration with respect to a parameter	$\left\{\int_a^b f_n(\xi)\,\mathrm{d}\xi\right\}$	$\int_a^b F(z, \xi)\,\mathrm{d}\xi$
Inverse transform by integration	$f_n = \frac{1}{2\pi \mathrm{j}} \oint_C F(z)\, z^{n-1}\,\mathrm{d}z$	$F(z)$
Inverse transform by differentiation	$f_n = \frac{1}{n!} \frac{\mathrm{d}^n}{\mathrm{d}\zeta^n} F\left(\frac{1}{\zeta}\right)\Big\vert_{\zeta=0}$	$F(z)$
Product of sequences	$\{f_n g_n\}$	$\frac{1}{2\pi \mathrm{j}} \oint_C F(\zeta)\, G\left(\frac{z}{\zeta}\right) \zeta^{-1}\,\mathrm{d}\zeta$
Correlation	$\left\{\sum_{k=0}^{+\infty} f_k g_{k-n}\right\}$	$F(z)\, G\left(\frac{1}{z}\right)$
Autocorrelation	$\left\{\sum_{k=0}^{+\infty} f_k f_{k-n}\right\}$	$F(z)\, F\left(\frac{1}{z}\right)$
Sum	$\sum_{n=0}^{+\infty} f_n$	$\lim_{\substack{z \to 1+ \\ \mathrm{Im}\, z = 0}} F(z)$
Sum of squares	$\sum_{k=0}^{+\infty} f_k^2$	$\frac{1}{2\pi \mathrm{j}} \oint_C F(\zeta)\, F\left(\frac{1}{\zeta}\right) \zeta^{-1}\,\mathrm{d}\zeta$
Initial value	f_0	$\lim_{z \to \infty} F(z)$
Final value	$\lim_{n \to +\infty} f_n$	$\lim_{\substack{z \to 1+ \\ \mathrm{Im}\, z = 0}} (z - 1)\, F(z)$

7.2. List of $\mathscr{Z}$ Transforms

TABLE 7.2 Sequences and Transforms

Number	$\{f_n\}$	$F(z)$
1	$f_0 = 1, \quad f_n = 0, \quad n > 0$	1
2	$f_n = 1^n$	$\dfrac{z}{z-1}$
3	$(-1)^n$	$\dfrac{z}{z+1}$
4	$f_n = 1, \quad n = ik,$ $f_n = 0, \quad n \neq ik$	$\dfrac{z^i}{z^i - 1}$
5	n	$\dfrac{z}{(z-1)^2}$
6	n^2	$\dfrac{z(z+1)}{(z-1)^3}$
7	n^3	$\dfrac{z(z^2+4z+1)}{(z-1)^4}$
8	n^4	$\dfrac{z(z^3+11z^2+11z+1)}{(z-1)^5}$
9	n^5	$\dfrac{z(z^4+26z^3+66z^2+26z+1)}{(z-1)^6}$
10	n^{i-1}	$\dfrac{N_i(z)}{(z-1)^i}$
11	$(-1)^n n^{i-1}$	$\dfrac{(-1)^i N_i(-z)}{(z+1)^i}$
12	$\dbinom{n}{2} = \dfrac{n(n-1)}{2!}, \quad n \geqq 1$	$\dfrac{z}{(z-1)^3}$

TABLE 7.2 – *continued*

Number	$\{f_n\}$	$F(z)$
13	$\binom{n}{3} = \frac{n(n-1)(n-2)}{3!}, \quad n \geqq 2$	$\frac{z}{(z-1)^4}$
14	$\binom{n}{m} = \frac{n(n-1)\ldots(n-m+1)}{m!}, \quad n \geqq m-1$	$\frac{z}{(z-1)^{m+1}}$
15	$(-1)^n \binom{n}{m}, \quad n \geqq m-1$	$\frac{(-1)^m z}{(z+1)^{m+1}}$
16	$\binom{n+k}{m}, \quad k \leqq m$	$\frac{z^{k+1}}{(z+1)^{m+1}}$
17	a^n	$\frac{z}{z-a}$
18	a^{n+1}	$\frac{za}{z-a}$
19	a^{n-1}	$\frac{1}{z-a}$
20	$(-1)^n a^n$	$\frac{z}{z+a}$
21	$1 - a^n$	$\frac{z(1-a)}{(z-1)(z-a)}$
22	a^{2n}	$\frac{z}{z-a^2}$
23	$f_{2n} = a^n$, $f_{2n+1} = 0$	$\frac{z^2}{z^2-a}$
24	na^n	$\frac{za}{(z-a)^2}$

TABLE 7.2 – *continued*

Number	$\{f_n\}$	$F(z)$
25	n^2a^n	$\dfrac{az(a+z)}{(z-a)^3}$
26	n^3a^n	$\dfrac{az(z^2+4az+a^2)}{(z-a)^4}$
27	$n^{i-1}a^n$	$\dfrac{a^iN_i\left(\dfrac{z}{a}\right)}{(z-a)^i}$
28	$\dbinom{n}{2}a^n, \quad n \geqq 1$	$\dfrac{a^2z}{(z-a)^3}$
29	$\dbinom{n}{3}a^n, \quad n \geqq 2$	$\dfrac{a^3z}{(z-a)^4}$
30	$\dbinom{n}{m}a^n, \quad n \geqq m-1$	$\dfrac{a^mz}{(z-a)^{m+1}}$
31	$Aa^n + Bb^n$	$\dfrac{z^2(A+B)-z(Aa+Bb)}{(z-a)(z-b)}$
32	$\dfrac{1}{a-b}(a^{n+1}-b^{n+1})$	$\dfrac{z^2}{(z-a)(z-b)}$
33	$\mathrm{e}^{\alpha n}$	$\dfrac{z}{z-\mathrm{e}^{\alpha}}$
34	$\sinh \alpha n$	$\dfrac{z\sinh\alpha}{z^2-2z\cosh\alpha+1}$
35	$\cosh \alpha n$	$\dfrac{z(z-\cosh\alpha)}{z^2-2z\cosh\alpha+1}$
36	$\sinh(\alpha n+\varphi)$	$\dfrac{z[z\sinh\varphi+\sinh(\alpha-\varphi)]}{z^2-2z\cosh\alpha+1}$

TABLE 7.2 – *continued*

Number	$\{f_n\}$	$F(z)$
37	$\cosh(\alpha n + \varphi)$	$\dfrac{z[z\cosh\varphi - \cosh(\alpha - \varphi)]}{z^2 - 2z\cosh\alpha + 1}$
38	$a^n \sinh \alpha n$	$\dfrac{za\sinh\alpha}{z^2 - 2za\cosh\alpha + a^2}$
39	$a^n \cosh \alpha n$	$\dfrac{z(z - a\cosh\alpha)}{z^2 - 2za\cosh\alpha + a^2}$
40	$n \sinh \alpha n$	$\dfrac{z(z^2 - 1)\sinh\alpha}{(z^2 - 2z\cosh\alpha + 1)^2}$
41	$n \cosh \alpha n$	$\dfrac{z[(z^2 + 1)\cosh\alpha - 2z]}{(z^2 - 2z\cosh\alpha + 1)^2}$
42	$\sin \beta n$	$\dfrac{z\sin\beta}{z^2 - 2z\cos\beta + 1}$
43	$\cos \beta n$	$\dfrac{z(z - \cos\beta)}{z^2 - 2z\cos\beta + 1}$
44	$\sin(\beta n + \varphi)$	$\dfrac{z[z\sin\varphi + \sin(\beta - \varphi)]}{z^2 - 2z\cos\beta + 1}$
45	$\cos(\beta n + \varphi)$	$\dfrac{z[z\cos\varphi - \cos(\beta - \varphi)]}{z^2 - 2z\cos\beta + 1}$
46	$e^{\alpha n} \sin \beta n$	$\dfrac{z\,e^{\alpha}\sin\beta}{z^2 - 2z\,e^{\alpha}\cos\beta + e^{2\alpha}}$
47	$e^{\alpha n} \cos \beta n$	$\dfrac{z(z - e^{\alpha}\cos\beta)}{z^2 - 2z\,e^{\alpha}\cos\beta + e^{2\alpha}}$
48	$(-1)^n e^{\alpha n} \sin \beta n$	$\dfrac{-z\,e^{\alpha}\sin\beta}{z^2 + 2z\,e^{\alpha}\cos\beta + e^{2\alpha}}$

TABLE 7.2 – *continued*

Number	$\{f_n\}$	$F(z)$
49	$(-1)^n e^{\alpha n} \cos \beta n$	$\dfrac{z(z + e^{\alpha} \cos \beta)}{z^2 + 2z e^{\alpha} \cos \beta + e^{2\alpha}}$
50	$a^n \sin(\beta n + \varphi)$	$\dfrac{z[z \sin \varphi + a \sin(\beta - \varphi)]}{z^2 - 2za \cos \beta + a^2}$
51	$a^n \cos(\beta n + \varphi)$	$\dfrac{z[z \cos \varphi - a \cos(\beta - \varphi)]}{z^2 - 2za \cos \beta + a^2}$
52	$n \sin \beta n$	$\dfrac{z(z^2 - 1) \sin \beta}{(z^2 - 2z \cos \beta + 1)^2}$
53	$n \cos \beta n$	$\dfrac{z[(z^2 + 1) \cos \beta - 2z]}{(z^2 - 2z \cos \beta + 1)^2}$
54	$na^n \sin \beta n$	$\dfrac{az(z^2 - a^2) \sin \beta}{(z^2 - 2az \cos \beta + a^2)^2}$
55	$na^n \cos \beta n$	$\dfrac{az[(z^2 + a^2) \cos \beta - 2az]}{(z^2 - 2az \cos \beta + a^2)^2}$
56	$T_n(x) = \cos(n \arccos x)$	$\dfrac{z(z - x)}{z^2 - 2zx + 1}$
57	$U_n(x) = \dfrac{\sin[(n + 1) \arccos x]}{\sqrt{(1 - x^2)}}$	$\dfrac{U_0(x) z^2}{z^2 - 2zx + 1}$
58	$a^n T_n(x)$	$\dfrac{z(z - ax)}{z^2 - 2axz + a^2}$
59	$a^n U_n(x)$	$\dfrac{U_0(x) z^2}{z^2 - 2axz + a^2}$
60	$\dfrac{1^{n-1}}{n}, \quad n \geqq 1$	$\ln \dfrac{z}{z - 1}$

TABLE 7.2 – *continued*

Number	$\{f_n\}$	$F(z)$
61	$\dfrac{(-1)^{n-1}}{n}, \quad n \geqq 1$	$\ln\left(1 + \dfrac{1}{z}\right)$
62	$\dfrac{a^{n-1}}{n}, \quad n \geqq 1$	$\dfrac{1}{a} \ln \dfrac{z}{z-a}$
63	$1^{n-1} \dfrac{\cos \beta n}{n}, \quad n \geqq 1$	$\ln \dfrac{z}{\sqrt{(z^2 - 2z\cos\beta + 1)}}$
64	$\dfrac{\sin \beta n}{n}$	$\arctan \dfrac{\sin\beta}{z - \cos\beta}$
65	$(-1)^{n-1} \dfrac{\cos \beta n}{n}, \quad n \geqq 1$	$\ln \dfrac{\sqrt{(z^2 + 2z\cos\beta + 1)}}{z}$
66	$(-1)^{n-1} \dfrac{\sin \beta n}{n}, \quad n \geqq 1$	$\arctan \dfrac{\sin\beta}{z + \cos\beta}$
67	$\dfrac{a^{2n-1}}{2n-1}, \quad n \geqq 1$	$\dfrac{1}{\sqrt{z}} \operatorname{arctanh} \dfrac{a}{\sqrt{z}}$
68	$\dfrac{\cos \beta(2n-1)}{2n-1}, \quad n \geqq 1$	$\dfrac{1}{4\sqrt{z}} \ln \dfrac{z + 2(\sqrt{z})\cos\beta + 1}{z - 2(\sqrt{z})\cos\beta + 1}$
69	$\dfrac{\sin \beta(2n-1)}{2n-1}, \quad n \geqq 1$	$\dfrac{1}{2\sqrt{z}} \arctan \dfrac{2(\sqrt{z})\sin\beta}{z-1}$
70	$\dfrac{a^n}{n!}$	$e^{a/z}$
71	$\dfrac{n+1}{n!} a^n$	$\left(1 + \dfrac{a}{z}\right) e^{a/z}$
72	$\dfrac{\sin \beta n}{n!}$	$e^{(\cos\beta)/z} \sin \dfrac{\sin\beta}{z}$

TABLE 7.2 – *continued*

Number	$\{f_n\}$	$F(z)$
73	$\dfrac{\cos \beta n}{n!}$	$e^{(\cos\beta)/z} \cos \dfrac{\sin\beta}{z}$
74	$\dfrac{a^n \sin \beta n}{n!}$	$e^{(a\cos\beta)/z} \sin \dfrac{a\sin\beta}{z}$
75	$\dfrac{a^n \cos \beta n}{n!}$	$e^{(a\cos\beta)/z} \cos \dfrac{a\sin\beta}{z}$
76	$\binom{k}{n}, \quad n \leqq k$	$\left(1 + \dfrac{1}{z}\right)^k$
77	$\dfrac{(-1)^n}{(2n+1)!}$	$(\sqrt{z}) \sin \dfrac{1}{\sqrt{z}}$
78	$\dfrac{(-1)^n}{(2n)!}$	$\cos \dfrac{1}{\sqrt{z}}$
79	$\dfrac{1}{(2n+1)!}$	$(\sqrt{z}) \sinh \dfrac{1}{\sqrt{z}}$
80	$\dfrac{1}{2n!}$	$\cosh \dfrac{1}{\sqrt{z}}$
81	$\dfrac{a^n}{(2n+1)!}$	$\left(\sqrt{\dfrac{z}{a}}\right) \sinh \sqrt{\dfrac{a}{z}}$
82	$\dfrac{a^n}{2n!}$	$\cosh \sqrt{\dfrac{a}{z}}$
83	$\dfrac{B_n}{n!}$, B_n are Bernoulli numbers	$\dfrac{1}{z(e^{1/z} - 1)}$
84	$\dfrac{B_n(x)}{n!} = \sum_{k=0}^{n} \binom{n}{k} \dfrac{B_k}{n!} x^{n-k}$	$\dfrac{e^{x/z}}{e^{1/z} - 1}$

TABLE 7.2 – *continued*

Number	$\{f_n\}$	$F(z)$
85	$\frac{E_n}{n!}$, E_n are Euler numbers	$\frac{1}{\cosh \frac{1}{z}}$
86	$L_n(x) = \sum_{k=0}^{n} \binom{n}{n-k} \frac{(-x)^k}{k!}$	$\frac{z}{z-1} e^{x/(1-z)}$
87	$\frac{H_n(x)}{n!} = \sum_{k=0}^{[n/2]} \frac{(-1)^k (2x)^{n-2k}}{k!\,(n-2k)!}$	$e^{(2x-1/z)/z}$
88	$P_n(x) = 2^{-n} \sum_{k=0}^{[n/2]} (-1)^k \binom{n}{k} \binom{2(n-k)}{n} x^{n-2k}$	$\frac{1}{\sqrt{(z^2 - 2xz + 1)}}$

7.3. Polynomials $N_i(z)$

TABLE 7.3 Recurrent Relations

$$N_i(z) = c_{i,0}z^i + c_{i,1}z^{i-1} + \ldots + c_{i,v}z^{i-v} + \ldots + c_{i,i-1}z + c_{i,i}$$

$$N_{i+1}(z) = iz\, N_i(z) - (z^2 - z)\frac{\mathrm{d}}{\mathrm{d}z} N_i(z), \qquad i \geqq 1$$

$$N_1(z) = z$$

$$N_{i+1}(1) = \sum_{v=0}^{i+1} c_{i+1,v} = i \sum_{v=0}^{i} c_{i,v} = i\, N_i(1)$$

$$c_{i,0} = c_{i,i} = 0$$

$$c_{i,1} = c_{i,i-1} = 1$$

$$c_{i+1,v} = (i + 1 - v)\, c_{i,v-1} + vc_{i,v}, \qquad i \geqq 1$$

$$c_{1,0} = 1$$

$$c_{1,1} = 0$$

TABLE 7.4 Coefficients

v	i									
	1	2	3	4	5	6	7	8	9	10
0	1	0	0	0	0	0	0	0	0	0
1	0	1	1	1	1	1	1	1	1	1
2		0	1	4	11	26	57	120	247	502
3			0	1	11	66	302	1 191	4 293	14 608
4				0	1	26	302	2 416	15 619	88 234
5					0	1	57	1 191	15 619	156 190
6						0	1	120	4 293	88 234
7							0	1	247	14 608
8								0	1	502
9									0	1
10										0

TABLE 7.5 Zeros

μ	i	
	1	2
1	0	0

μ	i	
	3	4
1	0	0
2	−1	$-0.267\,949\,192\,431 \,.\, 10^{0}$
3		$-0.373\,205\,080\,757 \,.\, 10^{1}$

TABLE 7.5 – *continued*

μ	i = 5	i = 6
1	0	0
2	$-0.101\,020\,514\,434 \,.\, 10^{0}$	$-0.430\,962\,882\,033 \,.\, 10^{-1}$
3	-1	$-0.430\,575\,347\,100 \,.\, 10^{0}$
4	$-0.989\,897\,948\,557 \,.\, 10^{1}$	$-0.232\,247\,388\,694 \,.\, 10^{1}$
5		$-0.232\,038\,544\,777 \,.\, 10^{2}$

μ	i = 7	i = 8
1	0	0
2	$-0.195\,242\,426\,904 \,.\, 10^{-1}$	$-0.914\,869\,480\,961 \,.\, 10^{-2}$
3	$-0.220\,170\,761\,011 \,.\, 10^{0}$	$-0.122\,554\,615\,192 \,.\, 10^{0}$
4	-1	$-0.535\,280\,430\,796 \,.\, 10^{0}$
5	$-0.454\,192\,916\,174 \,.\, 10^{1}$	$-0.186\,817\,963\,532 \,.\, 10^{1}$
6	$-0.512\,183\,758\,345 \,.\, 10^{2}$	$-0.815\,962\,743\,166 \,.\, 10^{1}$
7		$-0.109\,305\,209\,192 \,.\, 10^{3}$

μ	i = 9	i = 10
1	0	0
2	$-0.437\,615\,764\,594 \,.\, 10^{-2}$	$-0.212\,130\,690\,318 \,.\, 10^{-2}$
3	$-0.716\,504\,526\,689 \,.\, 10^{-1}$	$-0.432\,226\,085\,405 \,.\, 10^{-1}$
4	$-0.318\,709\,398\,786 \,.\, 10^{0}$	$-0.201\,750\,520\,193 \,.\, 10^{0}$
5	-1	$-0.607\,997\,389\,169 \,.\, 10^{0}$
6	$-0.313\,765\,456\,469 \,.\, 10^{1}$	$-0.231\,360\,399\,978 \,.\, 10^{2}$
7	$-0.139\,566\,456\,492 \,.\, 10^{2}$	$-0.164\,474\,390\,485 \,.\, 10^{1}$
8	$-0.228\,510\,963\,477 \,.\, 10^{3}$	$-0.495\,661\,671\,178 \,.\, 10^{1}$
9		$-0.471\,407\,507\,561 \,.\, 10^{3}$

7.4. Proofs of Theorems

PROOF OF THEOREM 1. The application of the Abel theorem to series (1.1) yields directly Theorem 1.

PROOF OF THEOREM 2. By the Weierstrass comparison criterion the series $\sum_{n=0}^{+\infty} f_n(\xi) z^{-n}$ converges uniformly with respect to ξ in the interval $[a, b]$ if there exists a convergent series with positive terms $\sum_{n=0}^{+\infty} c_n$ such that we have

$$|f_n(\xi) z^{-n}| \leqq c_n \tag{7.1}$$

for all n and all ξ from the interval $[a, b]$. Theorem 1 implies that it suffices to prove relation (7.1) for $z = z_0$. If the series $\sum_{n=0}^{+\infty} c_n$ is convergent, then the series $M \sum_{n=0}^{+\infty} c_n$, $M > 0$, is also convergent. Therefore, we choose as majorant the series $M \sum_{n=0}^{+\infty} q^n$, $0 < q < 1$. We obtain

$$|f_n(\xi) z_0^{-n}| \leqq Mq^n$$

and upon modification

$$|f_n(\xi) \leqq M(q|z_0|)^n < M|z_0|^n . \tag{7.2}$$

The uniform convergence of the series is not affected by dropping a finite number of terms. Therefore, it is sufficient if relation (7.2) is valid only for $n \geqq n_0 \geqq 0$.

PROOF OF THEOREM 3a. Theorem 3a follows from Theorem 2 for constant ξ.

PROOF OF THEOREM 3b. By Theorem 3a every sequence of exponential type is $\mathscr{Z}$ transformable. To prove the necessary condition assume that $\mathscr{Z}\{f_n\} = F(z)$ holds for $|z| > 1/R$. By the theorem on the inverse transform (Theorem 19) we have, for $n \geqq 0$, the relation

$$f_n = \frac{1}{2\pi j} \oint_C F(z)\, z^{n-1}\, dz ,$$

where integration is performed along the circle C given by the relation $z = \varrho e^{j\varphi}$, $\varrho > 1/R$, $0 \leqq \varphi \leqq 2\pi$. By the theorem on the estimate of an integral we have, for all $n \geqq 0$,

$$|f_n| = \frac{1}{2\pi} \left| \oint_C F(z)\, z^{n-1}\, dz \right| \leqq M\varrho^n , \tag{7.3}$$

where M is the maximum of the modulus of the function $F(z)$ on the circle C. If we put $\ln \varrho = s_0$ in (7.3), we obtain

$$|f_n| \leqq M e_0^{ns} ,$$

which is the definitorical relation of a sequence of the exponential type.

PROOFS OF THEOREMS 4a, 4b. Both theorems are the direct consequence of the theorem on the uniqueness of the Laurent series according to which the following holds for the series (1.1): *If the series* (1.1) *is convergent for* $|z| > 1/R$, *then the function* $F(z)$ *is regular in this region and the series* (1.1) *is its Laurent expansion in the neighbourhood of infinity.*

PROOF OF THEOREM 5. For the convergence of the series

$$\sum_{n=0}^{+\infty} \sum_{i=0}^{l} c_i f_{i,n} z^{-n}$$

it is sufficient if $|z| > \max 1/R_i$, $i = 0, 1, 2, \ldots, l$, where R_i is the radius of convergence for the ith summand. Since the constants c_i are independent of m and l is finite, it is possible to exchange the order of the summations and we arrive at relation (2.1).

PROOF OF THEOREM 6. The definitorical relation of the $\mathscr{Z}$ transform can be divided into two parts:

$$F(z) = \sum_{n=0}^{+\infty} f_n z^{-n} = \sum_{n=k}^{+\infty} f_n z^{-n} + \sum_{n=0}^{k-1} f_n z^{-n} .$$

In the first sum on the right-hand side we put $n = n' + k$, and in the result we again write n instead of n':

$$F(z) = \sum_{n=0}^{+\infty} f_{n+k} z^{-n-k} + \sum_{n=0}^{k-1} f_n z^{-n} = z^{-k} \sum_{n=0}^{+\infty} f_{n+k} z^{-n} + \sum_{n=0}^{k-1} f_n z^{-n} .$$

Upon modification we obtain relation (2.7a).

Analogously, relation (2.7b) is also proved. The substitution $n = n' - k$ and the basic assumption of the one-sided transform $f_n = 0$ for $n < 0$ are used.

PROOF OF THEOREM 7a. Obviously, if $\mathscr{Z}\{g_n\}$ exists, then $\mathscr{Z}\{f_m\}$ also exists since $\{f_m\}$ contains only zero terms besides the terms of the sequence $\{g_n\}$. Thus, we have

$$\mathscr{Z}\{f_m\} = \sum_{m=0}^{+\infty} f_m z^{-m} = \sum_{n=0}^{+\infty} g_n z^{-ni}.$$

The last relation is convergent for $|z^i| > 1/R$, i.e. for $|z| > R^{-1/i}$.

PROOF OF THEOREM 7b. If $\mathscr{Z}\{f_m\} = F(z)$ exists, then $\mathscr{Z}\{g_n\}$ also exists since only zero terms of the sequence $\{f_m\}$ are omitted in the sequence $\{g_n\}$. We have

$$\mathscr{Z}\{g_n\} = \sum_{n=0}^{+\infty} g_n z^{-n} = \sum_{m=0}^{+\infty} f_m z^{-m/i} = \sum_{m=0}^{+\infty} f_m (z^{1/i})^{-m}$$

for $|z^{1/i}| > 1/R$, and thus for $|z| > R^{-i}$.

PROOF OF THEOREM 8. By the definition of the $\mathscr{Z}$ transform the transform of the sequence $\{\lambda^n f_n\}$ is

$$\mathscr{Z}\{\lambda^n f_n\} = \sum_{n=0}^{+\infty} \lambda^n f_n z^{-n} = \sum_{n=0}^{+\infty} f_n \left(\frac{z}{\lambda}\right)^{-n}.$$

The series is convergent for $|z/\lambda| > 1/R$, i.e. for all z in the region $|z| > |\lambda|/R$.

PROOF OF THEOREM 9. Let us have the product

$$F(z)\, G(z) = \sum_{n=0}^{+\infty} f_n z^{-n} \sum_{n=0}^{+\infty} g_n z^{-n}.$$

According to the Cauchy formula for the multiplication of absolutely convergent series the above relation can be written in the form

$$F(z)\, G(z) = \sum_{n=0}^{+\infty} \sum_{k=0}^{n} f_{n-k} g_k z^{-n}.$$

By the uniqueness theorem for the $\mathscr{Z}$ transform this series is the transform of the convolution of sequences $\{f_n\}$ and $\{g_n\}$, and it is convergent at least in the region $|z| > \max(1/R_1, 1/R_2)$.

PROOF OF THEOREM 10a. If the transform $\mathscr{Z}\{f_n\}$ exists, then the transform

$$\mathscr{Z}\{\Delta f_n\} = \mathscr{Z}\{f_{n+1}\} - \mathscr{Z}\{f_n\} = z[F(z) - f_0] - F(z)$$

also exists by the linearity and translation theorems. Upon modification we hence obtain relation (2.14). The same theorems imply that the radius of convergence remains unchanged.

PROOF OF THEOREM 10b. This theorem can be easily proved by the repeated application of Theorem 10a.

PROOF OF THEOREM 11. Let sequence $\{g_n\}$ be given by the formula

$$g_0 = 0,$$
$$g_n = 1 \qquad \text{for} \qquad n \geqq 1.$$

Its transform, for $|z| > 1$, is

$$G(z) = \frac{1}{z - 1}.$$

The sequence of partial sums can be represented by the convolution

$$\sum_{k=0}^{n-1} f_k = \sum_{k=0}^{n-1} f_k g_{n-k} = \sum_{k=0}^{n} f_k g_{n-k},$$

since $g_0 = 0$. With the aid of the theorem on convolutions we obtain relation (2.20) which converges for $|z| > \max(1, 1/R)$.

PROOF OF THEOREM 12. The series which defines $F(z)$ is a series of regular functions in the region $|z| > 1/R$. By Theorem 1, this series converges uniformly in every closed region $|z| \geqq R' > 1/R$, and it can be differentiated term by term according to the Weierstrass theorem. The newly obtained series is again uniformly convergent in the region $|z| > R'$, and it defines a function regular in the region $|z| > 1/R$. By the uniqueness theorem, this function is the unique transform of the sequence $\{nf_n\}$. Differentiation of the function $F(z)$ yields

$$\frac{\mathrm{d}}{\mathrm{d}z} F(z) = \frac{\mathrm{d}}{\mathrm{d}z} \sum_{n=0}^{+\infty} f_n z^{-n} = -\sum_{n=0}^{+\infty} n f_n z^{-n},$$

from which relation (2.22) follows by modification.

PROOF OF THEOREM 13. By Theorem 4a, the function $F(z)$ is a regular function in the region $|z| > 1/R$, and the series which defines it is uniformly convergent in a closed region lying inside the region $|z| > 1/R$. Consequently, it is possible to exchange the order of integration and summation. Under the assumption that the path of integration lies inside the region of regularity of the function $F(z)$, the integral is independent of the path.

By assumption, for $|\zeta| > 1/R$ we have

$$F(\zeta) = \sum_{n=0}^{+\infty} f_n \zeta^{-n}.$$

We multiply both series by ζ^{-1} and integrate from z to z_0:

$$\int_z^{z_0} F(\zeta)\,\zeta^{-1}\,\mathrm{d}\zeta = \sum_{n=0}^{+\infty} f_n \int_z^{z_0} \zeta^{-n-1}\,\mathrm{d}\zeta = \left[\sum_{n=0}^{+\infty} \frac{f_n}{n} \zeta^{-n}\right]\Bigg|_z^{z_0}. \tag{7.4}$$

If the primitive function is regular at the point $z = \infty$, and if it vanishes at this point (by the uniqueness theorem this corresponds to the condition $f_0 = 0$), it is possible to pass to the limit in relation (7.4) for $z_0 \to \infty$ and we obtain (2.26).

PROOF OF THEOREM 14. The application of the theorem on the differentiation of uniformly convergent series to the definitorical relation gives directly Theorem 14.

PROOF OF THEOREM 15. The application of the theorem on the integration of uniformly convergent series to the definitorical relation yields Theorem 15.

PROOF OF THEOREM 16. The second Abel theorem applied to series (1.1) reads: *If* $\sum_{n=0}^{+\infty} f_n$ *is convergent, equality* (2.31) *holds under the assumption that* z^{-1} *approaches to the point* $z = 1$ *along an arbitrary path which lies between two chords of the unit circle passing through the point* $z = 1$. In the z plane, to these chords correspond arcs of circles which lie outside the unit circle and pass through the point $z = 1$. Thus, the path along which the point z has to approach the point $z = 1$ must lie outside the unit circle and in the region between the two arcs. It is obvious that this region includes the part of the real axis in the interval $[1, +\infty]$. This concludes the proof.

PROOF OF THEOREM 17. The function $F(z)$ is a regular function at the point $z = \infty$, thus $\lim_{z \to \infty} F(z)$ exists. The definition of the transform implies that this limit is equal to the initial value of the object function.

PROOF OF THEOREM 18. If $\mathscr{Z}\{f_n\} = F(z)$ exists, then the transform of the first difference

$$\mathscr{Z}\{\Delta f_n\} = (z - 1) F(z) - z f_0$$

also exists by Theorem 10a. However, for the transform of the first difference we also have

$$\mathscr{Z}\{\Delta f_n\} = \sum_{n=0}^{+\infty} \Delta f_n z^{-n} .$$

The comparison of the two relations implies the equality

$$(z - 1) F(z) = z f_0 + \sum_{n=0}^{+\infty} \Delta f_n z^{-n} .$$

By the assumption of the theorem, $\lim_{n \to +\infty} f_n = f_0 + \sum_{n=0}^{+\infty} \Delta f_n$ exists. Consequently, $\lim_{\substack{z \to 1+ \\ \operatorname{Im} z = 0}} (z - 1) F(z)$ also exists, which proves the theorem.

PROOF OF THEOREM 19. It is possible to expand the function $F(z)$ in the neighbourhood of infinity into the Laurent series $\sum_{n=-\infty}^{+\infty} f_n z^{-n}$. However, the obtained expansion is constituted only by the regular part of the Laurent expansion in the neighbourhood of the point $z = \infty$ in consequence of the regularity of the function $F(z)$ for $|z| > 1/R$, i.e. it is given by a series in the negative powers of the variable z. For the sequence of coefficients of this series we have relation (2.34). By Theorem 4b, this sequence is the only sequence whose transform is exactly $F(z)$.

PROOF OF THEOREM 20. The function $F(z)$ is regular at the point $z = \infty$. Thus, it is possible to expand the function $F(1/\zeta)$, regular at the point $\zeta = 0$, into a Taylor series the coefficients of which are given by relation (2.39). The uniqueness theorem implies that the sequence of coefficients of the Taylor expansion is the only sequence whose transform is $F(z)$.

PROOF OF THEOREM 21. If the sequences $\{f_n\}$ and $\{g_n\}$ are $\mathscr{Z}$ transformable, they are of the exponential type by Theorem 3b. Then the sequence $\{f_n g_n\}$ is also of the exponential type and its transform is

$$\mathscr{Z}\{f_n g_n\} = \sum_{n=0}^{+\infty} f_n g_n z^{-n} \tag{7.5}$$

for $|z| > 1/(R_1 R_2)$. This proves the first assertion of Theorem 21.

By Theorem 19 we have

$$f_n = \frac{1}{2\pi \mathrm{j}} \oint_C F(\zeta)\, \zeta^{n-1}\, \mathrm{d}\zeta\,,$$

where we integrate along the circle C given by the relation $\zeta = \varrho \mathrm{e}^{\mathrm{j}\varphi}$, $\varrho > 1/R_1$, $0 \leqq \varphi \leqq 2\pi$. Upon substitution into (7.5) we have

$$\mathscr{Z}\{f_n g_n\} = \frac{1}{2\pi \mathrm{j}} \sum_{n=0}^{+\infty} \oint_C F(\zeta)\, \zeta^{n-1}\, \mathrm{d}\zeta\, g_n z^{-n} \tag{7.6}$$

for $|z| > 1/(R_1 R_2)$ and $|\zeta| > 1/R_1$. We construct the series $\sum_{n=0}^{+\infty} g_n \zeta^n z^{-n}$.

By the theorem on the similarity of transforms we have, for $|z| > |\zeta|/R_2$,

$$\sum_{n=0}^{+\infty} g_n \left(\frac{z}{\zeta}\right)^{-n} = G\left(\frac{z}{\zeta}\right). \tag{7.7}$$

This series is absolutely convergent for $|\zeta| < |z| R_2$ (z fixed), and in every closed region lying inside this region it converges uniformly to ζ.

The function $F(\zeta)$ is regular for $|\zeta| > 1/R_1$. Consequently, $F(\zeta)/\zeta$ is also a regular function in the same region and it is bounded on the circle C. Thus, the series

$$\frac{F(\zeta)}{\zeta} \sum_{n=0}^{+\infty} g_n \left(\frac{z}{\zeta}\right)^{-n} = \frac{F(\zeta)}{\zeta} G\left(\frac{z}{\zeta}\right) \tag{7.8}$$

converges uniformly to ζ. Let us multiply relation (7.8) by the value $1/(2\pi j)$ and integrate along the circle C:

$$\frac{1}{2\pi j} \oint_C \frac{F(\zeta)}{\zeta} G\left(\frac{z}{\zeta}\right) d\zeta = \frac{1}{2\pi j} \oint_C \frac{F(\zeta)}{\zeta} \sum_{n=0}^{+\infty} g_n \left(\frac{z}{\zeta}\right)^{-n} d\zeta\,.$$

The exchange of integration and summation yields relation (7.6). Let us review the conditions for z and ζ at this point:

Relation (7.5), and thus also relation (7.6), implies a condition for $|z|$, namely $|z| > 1/(R_1 R_2)$. Relations (7.6) and (7.7) yield a condition for $|\zeta|$, $1/R_1 < \zeta < |z|/R_2$. Thus, we integrate in the positive sense along a circle inside which lie all the singular points of the function $F(\zeta)$ and outside which lie all the singular points of the function $G(z/\zeta)$.

PROOF OF THEOREM 22. A function $F_{\mathrm{II}}(z)$ which is regular in the annulus $1/R_+ < |z| < R_-$ can be expanded into the Laurent series $\sum_{n=-\infty}^{+\infty} f_n z^{-n}$, in which the sequence of coefficients of the expansion is given by relation (2.58). In consequence of the uniqueness of the Laurent expansion this sequence is the only expansion of the series $F_{\mathrm{II}}(z)$.

PROOF OF THEOREM 23. By the assumptions of the theorem, the function $F_{\mathrm{II}}(z)$ which is regular in the annulus $1/R_+ < |z| < R_-$ is decomposed into the part $F_+(z)$, which is regular at the point $z = \infty$, and the part $F_-(z)$, which is regular at the point $z = 0$. Consequently, it is possible to expand the functions $F_+(1/\zeta)$ and $F_-(z)$ into a Taylor series whose coefficients are given by relations (2.63a) and (2.63b). The uniqueness theorem applied to the partial series implies that the obtained sequence of coefficients is the only sequence whose transform is $F_{II}(z)$.

PROOF OF THEOREM 24. The proof is performed with the aid of the Poisson summation formula [3] which reads:

Let $\varphi(x)$ be a piecewise continuous function with bounded variation in the interval $(-\infty, +\infty)$ and let it satisfy, for sufficiently large $A > 0$, $B > 0$, at least one of the following conditions in the intervals $(-\infty, -B)$, $(A, +\infty)$:

1. *$\varphi(x)$ is monotonic and absolutely integrable,*
2. *$\varphi(x)$ is integrable and has an absolutely integrable derivative.*

Further, let the series $\sum_{k=-\infty}^{+\infty} \psi(k)$ and $\sum_{n=-\infty}^{+\infty} \varphi(n)$ be convergent, where

$$\psi(k) = \int_{-\infty}^{+\infty} \varphi(x)\, e^{2\pi jkx}\, dx\,, \quad k \text{ integer}\,. \tag{7.9}$$

Then we have

$$\sum_{k=-\infty}^{+\infty} \psi(k) = \sum_{n=-\infty}^{+\infty} \varphi(n)\,. \tag{7.10}$$

A series $\sum_{k=-\infty}^{+\infty} \Phi(k)$ is called convergent if $\lim_{n\to+\infty} \sum_{k=-n}^{n} \Phi(k)$ exists. The integral in relation (7.9) is understood in the sense of its principal value.

Now, we proceed to the proof proper of Theorem 24. In relation (4.38) we put $t = Tx$ and obtain

$$F(p) = T \int_{0}^{+\infty} f(Tx)\, e^{-pTx}\, dx\,. \tag{7.11}$$

We define the function in relation (7.9) as follows:

$$\begin{aligned}
\varphi(x) &= f(Tx)\, e^{-pTx} && \text{for} \quad x > 0\,,\\
\varphi(x) &= 0 && \text{for} \quad x < 0\,,\\
\varphi(x) &= \frac{f(0)}{2} && \text{for} \quad x = 0\,.
\end{aligned}$$

Relation (7.9) can then be written in the form

$$\psi(k) = \int_0^{+\infty} f(Tx)\,\mathrm{e}^{-xT(p-2\pi jk/T)}\,\mathrm{d}x\,.$$

Comparison with relation (7.11) yields

$$\psi(k) = \frac{1}{T} F_{\mathrm{L}}\left(p - 2\pi\mathrm{j}\,\frac{k}{T}\right).$$

The validity of $f(t) = 0$ for $t < 0$ implies, from relation (7.10), the equality

$$\frac{1}{T} \sum_{k=-\infty}^{+\infty} F_{\mathrm{L}}\left(p - 2\pi\mathrm{j}\,\frac{k}{T}\right) = \frac{f_0}{2} + \sum_{n=1}^{+\infty} f(nT)\,\mathrm{e}^{-npT}\,.$$

Adding $f(0)/2$ to both sides and changing the subscript k to $-k$ on the left-hand side, we obtain (4.39).

We perform the analysis of the conditions of relation (4.39). By assumption the function $f(t)$ is differentiable for $t > 0$. Thus, even the function $\varphi(x) = f(xT)\,\mathrm{e}^{-pxT}$ is differentiable for $t > 0$ and, consequently, also continuous for $t > 0$. Besides that, if for the function $f(t)$ the integral

$$\int_0^{+\infty} |f'(t)|\,\mathrm{e}^{-p_0 t}\,\mathrm{d}t \tag{7.12}$$

is convergent, where p_0 is a real number, then $\varphi(x) = f(xT)\,\mathrm{e}^{-pxT}$ is a function with bounded variation. The function $\varphi(x) = f(xT)\,\mathrm{e}^{-pxT}$ does not generally satisfy the first condition of the Poisson summation formula (monotonicity of $\varphi(x)$). However, if the function $f(t)$ satisfies the convergence condition (7.12), then by the theorem on the Laplace transform of a derivative the integral of relation (4.38) also exists and, moreover, it converges absolutely for $\mathrm{Re}\,p > p_0$, the limit $\lim_{t\to 0} f(t) = f_0$ exists, and $f(t)$ is a function of the exponential type.

References

[1] Ackroyd, M. H.: *Digital Filters.* Computers in Medicine Series. Butterworths, London 1973.

[2] Antoniou, A.: *Digital Filters: Analysis and Design.* McGraw–Hill, New York 1979.

[3] Bochner, S.: *Vorlesungen über Fouriersche Integrale.* Akademische Verlagsgesellschaft, Leipzig 1932.

[4] Bogert, B. P., Healy, M. J. and Tukey, J. W.: The frequency analysis of time series for echoes: cepstrum, pseudo-autocovariance, cross-cepstrum and saphe cracking. In: *Proc. Symp. Time-Series Analysis,* M. Rosenblatt Ed., J. Wiley, New York 1963, 209–243.

[5] Bogner, R. E. and Constantinides, A. G.: *Introduction to Digital Filtering.* J. Wiley, London 1975.

[6] Burrus, C. S. and Parks, T. W.: Time domain design of recursive digital filters. *IEEE Trans. on Audio and Electroacoustics* **AU-18**, 1970, No. 2, 137–141.

[7] Cadzow, J. A.: *Discrete Time Systems.* Prentice-Hall, New Jersey 1973.

[8] Childers, D. G., Skinner, D. P. and Kemerait, R. C.: The cepstrum: A guide to processing. *Proc. IEEE,* **65**, No. 10, October 1977, 1428–1443.

[9] Childers, D. G.: *Modern Spectrum Analysis.* IEEE Press, New York 1978.

[10] Čížek, V.: *Discrete Fourier Transforms and Their Applications.* Adam Hilger, Bristol 1986.

[11] Doetsch, G.: *Handbuch der Laplace-Transformation.* Birkhäuser Verlag, Basel 1950.

[12] Doetsch, G.: *Anleitung zum praktischen Gebrauch der Laplace-Transformation.* Chap. 4. Oldenbourg Verlag, München 1961.

[13] Gold, B. and Rader, Ch. M.: *Digital Processing of Signals.* McGraw–Hill, New York 1969.

[14] James, H. M., Nichols, N. B. and Phillips, R. S.: *Theory of Servomechanisms*. MIT Lab. Series 25, New York 1947.

[15] Jury, E. I.: *Sampled-Data Control Systems*. J. Wiley, New York 1958.

[16] Jury, E. I.: *Theory and Application of the $\mathscr{Z}$-Transform Method*. J. Wiley, New York 1964.

[17] Jutzi, K. H.: Frequenzbestimmung durch Signalabtastung. *Frequenz* **27**, Nov. 1973, 11, 301 – 311.

[18] Khovanskiy, A. N.: *Application of continued fractions and their generalizations to problems of approximate analysis*. (In Russian.) Gosudarstvennoe izdatelstvo tekhniko-teoreticheskoi literatury, Moscow 1956.

[19] Klein, W.: *Finite Systemtheorie*. Teubner, Stuttgart 1976.

[20] Kress, D.: *Theoretische Grundlagen der Signal- und Informationsübertragung*. Akademie-Verlag, Berlin 1977.

[21] Kulhánek, O.: *Introduction to Digital Filtering in Geophysics*. Elsevier, Amsterdam 1976.

[22] Lacroix, A.: *Digital Filter. Eine Einführung in zeitdiskrete Signale und Systeme*. Oldenbourg Verlag, München 1980.

[23] Lange, F. H.: *Signale und Systeme*. Verlag Technik, Berlin 1965.

[24] Lücker, R.: *Grundlagen digitaler Filter*. Springer Verlag, Berlin 1980.

[25] Markel, J. D. and Gray, A. H.: *Linear Prediction of Speech*. Springer Verlag, Berlin 1976.

[26] Mason, S. J. and Zimmermann, H. J.: *Electronic Circuits, Signals and Systems*. J. Wiley, New York 1960.

[27] Mitra, S. K. and Burrus, Ch. S.: A simple efficient method for the analysis of structures of digital and analog systems. *A. E. Ü.* **31**, 1977, No. 1, 33 – 36.

[28] Neugebauer, H. J., Neunhäfer, H., Plešinger, A., Ruegg, J. C., Schick, R., Souriau, M., Stockl, H., Vích, R. and Zürn, W.: A comparison of different restitution methods using shake table experiments. *Zeischrift f. Geophysik*, **39**, 1973, 563 – 566.

[29] Oppenheim, A. V. and Schafer, R. W.: *Digital Signal Processing*. Prentice–Hall, New Jersey 1975.

[30] Plešinger, A. and Vích, R.: On the identification of seismometric systems and the correction of recorded signals for identified transfer functions. *Zeitschrift f. Geophysik*, **38**, 1972, 543 – 554.

[31] Rabiner, L. R. and Rader, Ch. M.: *Digital Signal Processing*. IEEE Press, New York 1972.

[32] Rabiner, L. R. and Gold, B.: *Theory and Application of Digital Signal Processing*. Prentice–Hall, New York 1975.

[33] Rabiner, L. R. and Schafer, R. W.: *Digital Processing of Speech Signals*. Prentice–Hall, New Jersey 1978.

[34] Robinson, E. A.: *Statistical Communication and Detection with Special Reference to Digital Data Processing of Radar and Seismic Signal*. Griffin, London 1967.

[35] Robinson, E. A. and Silvia, M. T.: *Digital Signal Processing and Time Series Analysis*. Holden–Day, San Francisco 1978.

[36] Schüssler, H. W.: *Digitale Systeme zur Signalverarbeitung*. Springer-Verlag, Berlin 1973.

[37] Steiglitz, K. and Dickinson, B.: Computation of the complex cepstrum by factorization of the $\mathscr{Z}$-transform. In: *IEEE Int. Conf. on ASSP* 1977, 723 – 726.

[38] Stoer, J.: *Einführung in die numerische Mathematik I*. Springer-Verlag, Berlin 1976.

[39] Štursa, J. and Vích, R.: Digital ladder filter as a real time approximation of partial differential equations. In: *Proc. of the 5th Colloquium on Microwave Comm.*, **II**, Akadémiai Kiadó, Budapest 1974, 299 – 308.

[40] Štursa, J. and Vích, R.: Problems of digital filter simulation. In: *Proc. of the Sixth Summer Symposium on Circuit Theory 1982*, Main Lectures, Inst. of Radio Engg. and Electronics, Prague 1982, 100 – 115.

[41] Tou, J. T.: *Digital and Sampled-Data Control Systems*. McGraw–Hill, New York 1959.

[42] Treitel, S. and Robinson, E. A.: The stability of digital filters. *Trans. IEEE on Geosci. Electr.* **GE-2**, 1964, 6 – 18.

[43] Tribolet, J. M. and Quartieri, T. F.: Computation of the complex cepstrum. In: *Programs for Digital Signal Processing*, IEEE Press 1979.

[44] Tzypkin, Ya. Z.: *Theory of linear impulse systems*. (In Russian.) Gosudarstvennoe izdatelstvo fiziko-matematicheskoi literatury, Moscow 1963.

[45] Unbehauen, R.: *Systemtheorie*. Oldenbourg Verlag, München 1971.

[46] Vích, R.: *$\mathscr{Z}$-Transformation, Theorie und Anwendung*. Verlag Technik, Berlin 1964.

[47] Vích, R.: Approximation in the time domain. In: *Proc. of the 3rd Colloquium on Microwave Comm.*, 1966, Akadémiai Kiadó, Budapest 1968, 359–365.

[48] Vích, R.: Selective properties of digital filters obtained by convolution approximation. *Electronics Letters*, **4**, 1968, 1–2.

[49] Vích, R.: Properties of digital bandpasses obtained by convolution approximation for zero frequency. *Electronics Letters*, **6**, 1970, 440–442.

[50] Vích, R.: Approximation in digital filter synthesis based on time-response invariance. *Electronics Letters*, **6**, 1970, 442–444.

[51] Vích, R.: Two methods for the construction of transfer functions of digital integrators. *Electronics Letters*, **7**, 1971, 422–425.

[52] Vích, R.: Bilateral deconvolution. *Zeitschrift f. Geophysik*, **39**, 1973, 557–562.

[53] Vích, R.: Matrix analysis and synthesis of multimesh digital filters. In: *Proc. of the 4th Summer School on Circuit Theory*, **1**, Inst. for Radio Engg. and Electronics, Prague 1974, 190–205.

[54] Vích, R.: True simulation of digital filters in the time and frequency domains. In: *Proc. of the 1981 European Conf. on Circuit Theory and Design*, Boite R., Dewilde P. editors, Delft University Press/North Holland, The Hague, 865–870.

[55] Vích, R.: Synthesis of digital polyphase networks. In: *Signal processing: Theories and Applications*, M. Kunt and F. de Coulon editors, North Holland 1980, 245–250.

[56] Vích, R.: Multimesh digital filter realization of the Goertzel algorithm for sliding spectrum analysis. In: *Proc. of the 1980 European Conf. on Circuit Theory and Design*, **2**, Warsaw 1980, 456–461.

[57] Vích, R.: Cepstral analysis by factorization of the signal $\mathscr{Z}$-transform. In: *Proc. of the 1983 European Conf. on Circuit Theory and Design*, Stuttgart 1983, 266–268.

[58] Vích, R.: Generalized synthesis of recursive digital filters in the frequency domain by linear programming. *EUSIPCO-83 Signal Processing: Theories and Applications*. H. W. Schüssler editor, North Holland 1983, 49–52.

[59] Woschni, E. G.: *Informationstechnik*. Verlag Technik, Berlin 1973.

[60] Wunsch, G.: *Systemanalyse*. Vol. 1, Verlag Technik, Berlin 1970.
[61] Wunsch, G.: *Systemtheorie der Informationstechnik*. Geest–Portig, Leipzig 1971.

Index